What Price the Countryside?

Phil Drabble

What Price the Countryside?

Michael Joseph

LONDON

First published in Great Britain by Michael Joseph Ltd
44 Bedford Square, London WC1
1985

British Library Cataloguing in Publication Data

Drabble, Phil
 What price the countryside?
 1. Nature conservation—Great Britain
 2. Landscape conservation—Great Britain
 I. Title
 333.76'16'0941   QH77.G7

ISBN 0–7181–2345–X

Photorypeset in Great Britain by
Input Typesetting Limited, London
Printed and bound by Billing & Son, Limited
London and Worcester

Contents

List of Illustrations

Acknowledgments

The author and publishers would like to thank the
following who provided illustrations:
Aerofilms, 1, 2, 3; Aquila Photographics, 4 (R. T. Mills),
8 (R. Siegal), 22 (A. W. Cundall); Richard Muir, 5, 6;
British Tourist Authority, 7, 11, 12, 13, 21; S. C. Porter,
10; Ron Duggins, 14; Robin Williams, 15, 20; David
Grewcock, 16, 17; *Sporting Gun*, 18; Chris Dawn/*Angling
Times*, 19.

Preface

The English countryside is still more varied and more beautiful than any land I know. It has been evolving and changing continuously since our primitive ancestors forsook hunting for farming and made clearings in the forest for their villages and crops.

Initially, such changes can have been no more than a gnatbite in the landscape because, in William the Conqueror's time, a population of only a million and a half was thinly spread across the countryside. Crops and settlements would scarcely have taken a slice from the cake. About forty times as many of us are now treading on each other's toes in the same area, and overpopulation, whether of mice or men, breeds inevitable aggression. There is now cut-throat competition for the right to work and play on the same patch. Advances in mechanised farming, intensive husbandry and chemical crop protection have accelerated slow change to a mad stampede that has destroyed more landscape in my time than in any previous generation

Increased spare time, whether enforced by unemployment or bestowed by 'labour saving' techniques, has mushroomed, till the leisure industry gets top priority because it is all that finds anything constructive for idle hands to do. Since most people are concentrated in towns, the resultant conflict between urban and rural interests, those who wish to make the countryside their playground or retain it as their workshop, has never been more bitter. The vast majority of both town and countryfolk are prepared to compromise, but dispute settled by sweet reason does not make the news. A tiny minority, the lunatic fringe of both camps, make matters worse by hitting the headlines.

However desirable the romantics believe their idealised Utopia to be, it is not practicable to stop the clock and freeze the countryside as lifeless as a trout in aspic. It will continue to evolve and change, whether they like it or not. Slumps and booms and strikes may one day be tamed, till agro-industrial lions lie down with sentimental lambs. But there is no way of stopping the present industrial revolution, with its silicon chips and automation creating more leisure which must be filled or folk will die of boredom. Easy transport has already brought the most unattainable solitudes no more than a canful of petrol away from every teeming city.

Flip the coin over and the unacceptable faces of Big Business and arrogant bureaucracy will leer out at you. Chemical farming, which has obliterated so much wildlife since the war, is here to stay. Either incentives or pressures will be needed to persuade or force the agro-chemical industry to do research on substances which are far more selective and far less persistent than some of the dangerous obscenities that are now sold to be deluged over the earth. Mechanised farming cannot have unlimited licence to bulldoze vast tracts of countryside into deserts of plough, from horizon to horizon, simply to bloat faceless financiers, in distant cities, who never even see the land they pay managers to farm.

Compromise between such divergent interests will not only cost money. It will mean substantial give-and-take of 'rights' and 'privileges', held dear by fanatics both sides of the fence. Above all, it will mean cutting the bureaucrats, who call the tune, down to their proper size. Trendy conservationists, whose hobby is farmer-bashing, and country folk, who earn a living on the land and hit back, are really both off target. The true culprits are the politicians, who too often have vested interests, and the bureaucrats who build empires by spawning the schemes that are doing so much damage to the interests of both sides.

I grew up on the edge of Staffordshire's Black Country with the freedom to roam over the land farmed by patients of my doctor father, and I learned to love country that had few conventional attractions. The landscape was aesthetically bleak and unfashionable, but goldfinches – or seven-coloured linnets, as we called them – still fed on the thistles which thrived on the worked-out pit banks. Snipe drummed over the 'swags' – or subsidence pools – and the rabbits in the hedge banks had not yet been afflicted by myxomatosis. I poached some of them, with the local miners – and caught others by invitation of the Squire, the last great landowner to survive in the area. As I grew up, I spent holidays with uncles in Wiltshire and East Anglia and I am now lucky enough to be my own boss, earning my living working with delightful people, amongst glorious scenery all over the country.

In the first part of this book I have tried to share the pleasures of the countryside I have enjoyed so much. Then I have examined some of the factors and people who will influence its future, for better or for worse. I am acutely aware that it is impossible to put a cash value on anything so precious, but I have chosen, as the title, *What Price the Countryside?*, to concentrate the mind upon what we should lose by its destruction.

P.D.

1

The Squires' Land

I am now lucky enough to live in a cottage on the edge of a wood, six miles from the nearest town and twelve from the nearest main line railway station. I earn my living in beautiful and often remote countryside, working with delightful countryfolk. As a priceless bonus, I work for myself and own no man as master.

It was not always like that because I was bred and born twenty miles away on the edge of Staffordshire's industrial Black Country, where my father was a family doctor with a mining practice which straddled the boundary between town and country, and I mis-spent the first twenty-three years of my working life in an engineering factory before escaping to earn my living with my pen. Now I live in the sort of surroundings I would cheerfully have given my eye teeth to have visited on holiday while I worked for conventional gaffers. So my ears have been assailed by the arguments of both sides and, having slept on both sides of their blankets, I have had plenty of scope to size them up.

Born in 1914 and growing up between the wars, I have been acutely aware of violent change for as long as I can remember. The local squires were patients of my father and during the lifetime of the three generations I knew, I witnessed the decline of their estate from six thousand acres to nothing, at a time when great estates all over the land suffered similar disintegration. The Old Squire was a copybook example of the autocratic landowner derided by the modern school of left-wing trendies, for he was proud of the very qualities they so dislike. He was as territorial as a badger and festooned the estate with notices, announcing that 'Trespassers Will Be Spiflocated'. As nobody knew what spiflocated meant – and his gamekeepers were also allergic to trespassers – it was rare to see a stranger on his land. He was addicted to blood sports and he hunted or shot about five days a week, in season, and sat as a magistrate on the Bench on the sixth, to lock up the poachers his keepers had caught. The seventh day was reserved for God, except that he called on his Head Keeper, on his way back from church, to check from the new exhibits dangling from the keeper's gallows that his men had not been wasting time but had trapped their quota of vermin.

Although the skyline was pocked with the slagheaps of coalmines, the estate was on the borders of two hunts and they were annually bidden – not invited! – to hold a joint meet on the Old Squire's birthday. Masters of Foxhounds knew their places in the social pecking order, so the Old Squire's word was law. If they had disobeyed they could have lost a vital six thousand acres of their country. The Old Squire was an avid hunting man himself and his wife was just as keen. Well into her eighties she could have made plenty of modern show-jumping wenches seem clumsy duffers by comparison. Lady Vernon, as she was known, by courtesy if not by right, rode side-saddle, ruthlessly heading the field. This did not always meet the approval of her husband, who could be heard roaring, 'Hold hard, dem you! You'll break your blardy neck!', not so much, perhaps, out of concern for her neck as for his own place at the head of the field. On his birthday meet, a good run after the fox was the last thing that he wanted. The whole local population turned out and the huntsman was ordered not to get one fox away for a long hunt, but to chivvy foxes about the park, from covert to covert, so that all the spectators could view a fox and feel that they had participated in the sport.

At conventional modern meets, visitors on foot see hounds and huntsmen arrive but, after hounds are cast into cover and find their first fox, they disappear over the skyline – and that is the last the humble foot followers usually see. Not so on the Old Squire's birthday. He was far more concerned that his tenants and their neighbours, following on their own flat feet, should enjoy themselves than that hounds and huntsmen should have a long run. And Masters of Hounds who did not cooperate would not have been invited back.

Paternalistic and patronising, the trendies will whinge – but it was men like the Old Squire who created the countryside whose passing they now mourn. Sport was his passion and his estate was patterned with small coppices and spinneys, cunningly sited at strategic points. Woodland was planted along ridges of high ground and underplanted with wild rhododendrons and laurel and bramble and snowberry, not to satisfy the taste of aesthetic highbrows or to make farming easier, but to provide hiding cover for pheasants and foxes.

On shooting days the Old Squire and his friends could stand in valley bottoms while gamekeepers and beaters flushed pheasants from one high cover to fly over their guns to the sanctuary of another. These really high birds, curving like boomerangs on the wind, were extremely hard to hit. They required skill enough to separate the men from the boys and most of the flatulent tycoons who form modern shooting syndicates would have slunk home with smoking hot gun barrels but nothing in the bag.

The park across the moat from the house was also planted with much smaller spinneys, each surrounded by a post-and-rail fence, to keep grazing cattle and horses away from poisonous rhododendrons. These gave cover and sanctuary to a few pheasants but had really been

planted more for beauty than for sport, and led the eye
away through purple vistas to hazy woodland on the
horizon. There was a tower, at a high point in the park,
erected to the memory of Admiral Vernon, known to his
contemporaries as Old Grog. His name had gone down in
the annals of history for winning the battle of Portobello,
but ratings in the navy still bless his memory because he
was the man responsible for their traditional ration of rum.
Neither were historically earth-shattering events and the
family who occupied the land there for the best part of a
thousand years probably left more mark by breeding
Diomed, the first winner of the Derby.

Nevertheless, it is to such unsung landowning families
that we owe so much. They fashioned the countryside that
our generation is in danger of frittering away for profit in
featureless prairie farms. These are now managed ruthlessly by the yardstick of the balance sheet on behalf of
impersonal pension funds and finance houses. In the Old
Squire's time, the whole estate was let to tenant farmers,
for the Vernon family had been content to be squires or
soldiers or sailors, or, when all else failed, the younger sons
had gone into the church, in days when parsons were
still respected. 'Trade' was a dirty word and money was
something which gentlemen did not discuss, so there was
always an agent to look after the sordid routines of rents
and wages, hirings and firings. Many of these agents were
about as straight as old sows' tails and they often made
more on the side than the landowner paid them.

But the fact that so much land was farmed by small
tenant farmers worked wonders for the beauty of the countryside. When I was growing up, ploughing and sowing
and reaping were all done by horses and the old adage
that no muck is as good for the land as the farmer's boot
was never truer. Farms of less than two hundred acres were
the norm, which was small enough for the farmer to walk
over his land so often that he not only knew every field but

almost every furrow. The least thing wrong was bound to catch his eye and was quickly remedied. His modern impersonal counterparts tell farm managers to tell foremen to tell their tractor drivers to spew foul chemicals on the land, killing friend as well as foe.

Labour in the Old Squire's time was cheap so that a man was set-on for every fifty acres and, since birth control was a matter of abstinence, families were large and the countryside well stocked.

Although the Old Squire would tolerate no strangers trespassing over his estate, tenants and neighbours were welcome to take short cuts to market or to church or school, so that footpaths spider-webbed between important landmarks. Good manners were still counted as a virtue so that nobody abused hospitality by going where he was not welcome. If a field was sown with corn, nobody but a lout would cross it – and those who did would soon be looking for another job or cottage! Only when corn was harvested could short cuts be resumed instead of following tortuous field boundaries.

My father-in-law, who was often a guest when the Old Squire went shooting, once caught the rough end of his quill pen on the unfounded suspicion of bad manners! The Old Squire was in the habit of sending a succinct postcard simply saying, 'Shooting Holly Bank Gorse' – or wherever – '10 AM Friday, if fine. A.L.V.'. A card arrived one Monday when my father-in-law was away from home. When no reply had been received on Wednesday, the Old Squire fired his second salvo, which arrived on Thursday. It said, quite simply, 'What is not worth thanking for, is not worth having. A.L.V.'. My father-in-law was also a tetchy man but, to their credit, they made it up.

The Holly Bank Gorse, where he had been invited to shoot, was another legacy to posterity created by the Old Squire, not to beautify the countryside but to improve his sport. Contrary to popular belief, foxes do not live much in holes or earths except at cubbing time, when vixens go to ground with their litters. For the rest of the year they often spend their day in the shelter of reed beds or osier plantations or standing corn; I have found them in the boles of pollard willows and even on open plough. One of their favourite sanctuaries, perhaps because of its inviolability, is the dense cover of gorse thickets. It was therefore common for landowners to plant covers, like the Old Squire's Holly Bank Gorse, and the fringe benefit for posterity – as well as foxes – was that the landscape was splashed with gold for ever after, since gorse is out of bloom only when kissing's out of fashion!

Good things soon end and, when the Old Squire died, greedy death duties, and perhaps rapacious agents, played hell with his estate. His son, The Squire, also a patient of my father, was untold kind to me as a boy and gave me the freedom of his park to go rabbiting with my dog and ferrets. There were still plenty of rabbits but the number of acres had shrunk to about a third, farm rents were low and, like so many great estates, the rot had set in, as ruthless as woodworm in church timbers.

The familiar landscape of my youth perished in a decade after the Second World War. Horses were replaced by tractors so that one man was needed to do the arable work of four. Cattle herds that had been milked by hand were siphoned direct to churns at the press of a button, and fields that had been a comfortable size to work with horses had their hedges bulldozed out to suit juggernaut machines. The changes to the country scene had little to do with greedy farmers or idle workers. We were starting a second Industrial Revolution as traumatic as the first, a hundred and fifty years ago. This time it was not steam engines and factories that concentrated rural labour in the towns. It was computers and mechanisation and earthmoving equipment which was tearing the heart out from our familiar fields.

9

It is difficult to appreciate just how fast mechanical techniques develop in time of war. The sudden vital need for aerodromes produced bulldozers and scrapers that could alter whole landscapes. Cost did not matter provided the aerodromes or ammunition dumps, camouflaged factories or barracks sprouted quickly enough.

After the war, such machines were sold off at knockdown prices at Government Surplus sales so that contractors could uproot farm trees or grub out hedges, fill in pools or make irrigation lakes that would have been quite uneconomic with pre-war equipment. Even at surplus prices, such tackle is too expensive to stand idle, so that there were soon contractors prepared to compete fiercely for business and farmers cashed-in by having their fields landscaped, not for amenity but to make mechanised agriculture even more profitable.

When The Squire died his son, the Young Squire, succeeded him – and he was as kind to me as his forebears had been. To put the timescale in perspective, I dedicated a book to him, five years ago, and he phoned to thank me on the night of publication. 'Hello, Phil,' he said. 'This is the Young Squire ringing to thank you for the book. I'll have you know I'm no longer as young as I was. I am well past the eighty mark.' He can give me a decade and, watching three generations of his family fight a hopeless rearguard action in defence of their estate, has also given me a bird's-eye view of the similar changes that have altered the whole face of rural England in my own lifespan.

By the time death duties had tapewormed into the shrivelled vitals of the Young Squire's land, he was left with about six hundred acres, which is less than the patch coped with by the average farm manager. Since his father and grandfather had let their farms, there was no way he could cash-in on what was left so he had to rely on an unearned income from controlled rents, with no chance of regaining possession of his own domain to farm it for himself.

His ancestors' involvement with sport had ensured a staff of keepers, implacably hostile to predators. Hawks and owls, stoats and weasels, badgers and hedgehogs had all fetched up dangling from the vermin pole. Even domestic cats could only rely on a one-way ticket, as the hank of multi-coloured pussy tails on the end of the gibbet testified. But the Young Squire was as keen on natural history as his ancestors had been on sport. I first met him in the woodland next to his house, where he was absorbed in watching the courtship flight of a lesser spotted woodpecker. The Old Squire's keepers would have shot it to put in a glass case. By the time the Young Squire arrived, Hannam, the last Head Keeper, had been slowed down by *anni Domini* – and perhaps disheartened because his new master announced that he liked to see his wildlife alive better than dead. The slackening off of predator control resulted in the return of unpersecuted badgers to the estate for the first time for generations, and I have described in *No Badgers in My Wood** how I saw my first wild badgers at Hilton with the Young Squire as my guide.

His reign has coincided with events which have revolutionised the whole countryside. Before the last war, huge tracts of land were still contained in great estates, linked by each other's boundaries in a chain which encircled many square miles of country. The landowners, whose ancestors had often lived there for centuries, as the squires of Hilton had, had resisted urban development on their acres and placed such a high priority on sport that a glorious mosaic of wood and coppices and game covers had made the country beautiful. The three generations of the squires' family I knew as I grew up were robbed of their heritage by taxation, and they saw farm mechanisation change the face of England more than earthquakes could have done.

---

* No Badgers in My Wood, Michael Joseph, 1979

When the Young Squire sold off land to pay his father's death duties the farms were bought by the tenants at knock-down prices, because if outsiders had bought the land they could not have terminated the leases to get vacant possession, and the rents were too low to give an economic return. While the old landowners had let the farms, the standard of farming had been reasonable because the only ground on which a farmer would be dispossessed was bad farming. When the tenants got possession, many of them indulged in what would be regarded as asset stripping in the world of industry and commerce. They chopped down woods and flogged the turf from old traditional pastures, for such high prices that many of them paid the purchase price of the land with the timber and turf they sold. So the lawns of jerry-built urban houses, as identical as match-boxes, were laid with the fertility stripped off land of the great estates.

The Young Squire retreated into his shell until the last straw forced him to sell the remnant of the heritage his ancestors had owned since Norman times. They drove the M6 motorway bang through the middle of the estate and built the Hilton Service Station within spitting distance of Hannam's cottage. The Old Squire must be turning in his grave, and cursing at every motorist who trespasses across his land, oblivious of the risks they run of imminent spiflocation. It was the end of an era.

2

Sheep Country

It is tempting to think that, however much the quality of cultivated countryside has been ruined by the decline of the great estates, hill country is as it has always been. That is one of the reasons folk who spend their lives in offices and factories yearn to escape to wide open spaces for their leisure. It also causes bitter conflicts about anything which threatens to change the timeless solace of the hills.

I have enjoyed nothing better than my work on the television series about sheepdog trials, *One Man and His Dog,* that has been transmitted annually for almost a decade. It is an international competition between the best – or most successful! – shepherds and dogs in the British Isles. One reason for my pleasure is that it so often entails working in the loveliest hill country in the land.

Coming from the cultivated Midlands, I needed no persuasion that it would be impossible to farm hundreds of thousands of acres of the hills without a dog. Slopes are so steep that no tractor could climb them, far less cultivate their soil. The ground is usually too boggy or rocky for any plough to cultivate while ruthless ravines and precipices spell certain doom to cattle. Nothing grows naturally but heather, bracken and tufts of wiry grass. The land is only good for grazing sheep, shooting grouse, growing trees, rock climbing or rambling. Only shepherding, forestry and shooting can provide a financially viable return.

Romantics who appreciate the wild scenery persuade

themselves that it has always been like that, but the fact is that, in prehistoric times, scrub woodland clothed most of the lower slopes, for trees of one species or another covered two-thirds of the land. Early settlers made clearings for villages and felled trees for their fires. Their descendants enlarged clearings to grow crops instead of relying solely on hunting and gathering wild fruits for their diet. Tracts of woodland, laid waste by fire, could soon be growing crops. Far from being as they always were, the uplands have never ceased to change. Rocks are the obvious subconscious symbols of immutable eternity so that wild and rocky scenery is particularly attractive to our generation, racked as it is by insecurity, with scant reason for faith in tomorrow. The soft colours of the heather that clothes the moorlands blend with the skyline to provide an inviting carpet, stretching to infinity. The hills offer hope of return to basic values.

When we devise the course for our television sheepdog trials we take this into account. New viewers may know nothing about the technicalities of competitive trialling and might grow bored before they caught on. So we take great trouble to select spectacular views that merge with the skyline. Scores of letters record viewers' delight at the tranquillity of this wild country and the heart-stopping grandeur of wide skies and perpetual purple heather. They leave us in no doubt about how deep would be their resentment if anyone altered what they regard as the birthright of their ancestors, so that one year I tried a tongue-in-cheek experiment.

Our trials ground in the Welsh hills was on a remote hill farm bordered by a mountain that was commonland. As its name implied, the local commoners had grazing rights and bought sheep which they turned out to feed on the common for the summer season. Most of us like something for nothing, and the commoners turned out more sheep than the common could sustain, so, when they had eaten off the good grass, they set about the hawthorn and mountain ash bushes, ring-barked the hollies and young oaks, and scoffed the heather and almost anything else that was not poisonous. It was a classic case of overgrazing and all that survived were acres of bracken that the sheep wouldn't graze. This thrived and grew so thick that its shade killed off any remaining grass or heather the sheep had missed. The overstocking was so great that the farmer on whose land we held the trials had been invaded by flocks of hungry sheep which had by then devastated the common. He was faced with the expense of erecting a fence, not to keep his own stock in, as is normal practice, but to keep the starving hordes of his neighbours out. Having lashed out so much, he decided to finish the job by eradicating the bracken his side of the fence and sowing with productive grass seed.

The result was spectacular. The whole common was a sea of waving bracken, just taking on the first golden tints of autumn. 'Our' side of the fence, the grass was green and thriving in the sun and rain and mountain air. White sheep, unsullied by the faintest tinge of urban smog, blossomed visibly into mutton as they grazed. The commoners' scraggy, half-starved brutes ranged yards for every mouthful. The camera looked along the taut wire-net boundary fence while I pointed out the devastation of overgrazing on one side, and the marvels of modern science that had made the desert bloom, on ours. I don't admire all the tricks of intensive husbandry – and I suspected some of the subsidies for the improvement had been filched by the taxman from my pocket, to add to the mutton mountain for which I should also be called on to fork out more hard-earned cash for 'intervention' – so my dead-pan description was a little tongue-in-cheek, and I confess I laid it on a bit, because I was interested to discover viewer reactions. I went rather over the top, describing the bracken as jaundiced and the grass, our side, as a lush, ovine banquet. In the main, countryfolk agreed but part of the urban audience could not cast off their rose-tinted specs to see reality. They complained bitterly about my description of glorious, autumn-tinted bracken as jaundiced, and they thought the farmer should be castigated – or worse! – for meddling with nature. The fact that the bracken – or 'fern' as some of them miscalled it – had replaced the overgrazed heather which had replaced the burned-out trees, escaped their notice. When the countryside changes – or evolves? – continuously over the centuries, it is a very subjective judgement to decide which stage was 'right' or 'natural'.

Whatever view you take, the fact remains that making a living in the hills grows more and more difficult. It is possible to calculate output in sheep to the acre on rich lowland pastures, but it is the number of acres required to keep a sheep which is the significant yardstick of upland productivity. For some years, hill farmers have been subsidised under the Common Agricultural Policy, so they have had every incentive to overgraze even their own land, whatever seeds it may sow for future disaster. Nobody can cheat the laws of supply and demand for ever and the signs are that hill subsidies will shrink at the same time that hill grazing will deteriorate. Add to this the increasing pressures for access to all open country and I fear that many farmers on marginal land will be forced to sell their Land Rovers and buy bikes, or revert to their hill ponies.

It is perfectly natural for folk who live in crowded cities to wish to escape for peace and solitude when work is done. Unemployment, as we know it, will never go away because its root cause is neither bad management, as the lefties claim, nor bloody-minded unions as the right profess. This time the steam engines that powered the pulley wheels of industry have been made redundant by silicon chips and computers and automated conveyor belts where one man, touching buttons, can accomplish more than a shop-stewardship of workers a generation ago. There is no need to be a sainted prophet to predict that one man will soon be able to turn out the work of ten. Either one will have to work to keep nine in idleness, or the work will have to be shared, to give all immeasurably more leisure than we have at present. Either machines will be masters and men will be slaves or men will be wise enough to exploit machines so that all can live fuller and pleasanter lives. In either case, the obvious fact is that people will have vastly more leisure, enforced or willingly planned, and a problem of the future will be to derive pleasure instead of boredom from the extra hours at our disposal.

The modern nostalgia for simple country things, with roots as deep as history, provides a clue. Against all predictions, our sheep dog programme nets an audience of over six million, which is often more than the Nine O'Clock News and several times as many as more popular chat shows, or funny men. This may be partly a reaction to the incessant sex, violence and politics the highbrow left-wing planners inflict upon us, but it also goes far deeper than that. We have grown so bored with the competitive rat race that there is deep nostalgia for the simple, wholesome life of our rural ancestors, however harsh it may have been in reality.

Watching the telepathic affinity between the shepherds and their dogs puts our technological revolution into perspective. Boffins with their computers, but no sheepdogs, cannot round up sheep from crags that are inaccessible to the most sophisticated machines. Men who live such solitary lives, out in all weathers as their primitive ancestors were, are immune to the sedition disgorged by militant shop stewards. Dogs which can run perhaps fifty or sixty miles a day, over apparently impossible terrain, put the effete pooches which parade at dog shows into perspective. The battle of wills between stubborn sheep and clever dogs has not changed since men graduated from hunting to herding. The whole ritual, played out by men

17

and dogs against such a dramatic backcloth fires the most sluggish urban mind. The urge to share such simplicity – at least in theory! – is almost overwhelming.

However glamorous or satisfying such simple lives may look on colour television, the fact is that earning the sparsest living in the hills offers little but hardship as a reward. Common Market subsidies, which encourage hill farmers to grow more lambs, have also inflated the price of meat that ends up in mutton mountains. The hollow bonanza which sired the surplus sheep is self-defeating, because farmers are seduced into overgrazing their land, as the common on our programme had been stripped of its fertility. As the subsidies are eventually shrivelled, by the law of supply and demand, the impossible task of fattening sheep on clapped-out pastures will bankrupt small men and cause depopulation of the uplands, leaving the spent soil to revert to jaundiced bracken. All it will then be good for is to grow a forest of softwoods which will prompt conservationists to break into yet another chorus of belly-aching brays.

Even at its best, life in the hill country has always been incredibly hard by urban standards, however romantic the scenery is at the height of summer or when autumn heather is at its richest. Grass cannot grow under rocks and stones so that, at slack seasons when labour was cheap, whole families spent days on end gathering up the rock and stone that volcanoes had spewed out or ice-age avalanches had squandered. It was no use gathering stone from one potential plot of grass to deposit it on another. Imaginative forebears may have erected a few cairns and monuments on mountain tops, to commemorate their dear departed, but proper farmers were more practical. They built stone walls of the dross they collected, either on the boundaries of their land or in a mosaic of small crofts around their homestead. When snows were deep and lambs were due, they could then bring down their flocks close to the house, where they could take proper care of them. The land from which they had cleared the stones would keep more sheep, when the flock returned, because God's air and rain could reach the soil the rocks had previously covered. The purpose of the multiplicity of small paddocks round the house is to enable the shepherd to fold his flock by constant rotation from one paddock to the next, without ever remaining on the stale ground long enough for parasites to complete their life cycle. Sheep are wormy creatures whose guts are infested with parasites, some of which are spilled out onto the ground to be ingested by the next sheep grazing there. Intensive husbandry, often practised with lowland flocks, allows extremely dense stocking because of higher quality pasture. Under such circumstances it is often necessary to inject the flock at least every three weeks to keep internal parasites in check. Such injections are a comparative novelty of modern husbandry and traditional sheep farmers coped with the problem by never allowing their sheep to graze the same ground for anything but short periods. They accomplished this, when the flock was out on the hill, by moving sheep to fresh ground every day when they visited the flock with their sheep dog. When the flock came down to the farmstead to lamb, the same object was achieved by folding from paddock to paddock. The important thing was that the circuit must not be completed until the life cycle of the parasites was broken, so that any left by the grazing flock, last time, would have died before sheep returned to graze the pasture again, so avoiding the risk of reinfestation. To be a good stockman it is also necessary to be a good naturalist, and the shepherds of old had forgotten more basic biology than plenty of boffins will ever know.

Sheep live in a matriarchal society, as we too often do, and the gaffer-sheep are normally the old ewes. These ewes are very territorial so that if when they are young they are carefully shepherded to ensure they do not stray away from the extensive home range of their owners' hill farm, and if they have their lambs there, they will regard it as their territory and not stray away thereafter – provided, of course, that there is sufficient keep to sustain them. The lambs they produce wll grow up to respect, or be 'hefted' by, the invisible boundaries of their mothers' home range so that the whole flock will remain on the hill or mountain, where they were bred, born and reared, without the need of an expensive boundary fence or wall. Such hefted flocks are sold, as assets, with a farm because fresh sheep brought in would wander unrestrained by conventional fencing and would need constant shepherding. But where land is over-grazed and keep is scarce, even hefted sheep will stray for miles in search of pastures new.

3

The Horse's Eye

A great excitement of my childhood was to be taken for a picnic on the horse's eye. It was an annual treat while we were staying with my uncle, who was the family doctor for the villagers of Bratton in Wiltshire, and his life and ours were about as different as chalk and cheese. Chalk, as it happens, was the greatest novelty to me because our soil was light and dusty. Dirt, the local miners called it, while the soil in Uncle Tom's garden at Bratton was almost as pallid and glutinous as the chalk of the downs above.

The downs ended on the escarpment along the Bratton-Westbury road and, cut into the hillside, which looked vertical from below, was the solid white form of the White Horse of Westbury, proud neck arched, tail flowing and eye flashing. From the road level, it looked quite a small eye and the idea of perching on it, far less having a picnic, was as preposterous as a fly perching on an incandescent jewel. But sometime during the stay I was always taken, to prove that facts can be stranger – and infinitely more exciting – than any fiction about defiance of gravity.

We set off after lunch, walking from uncle's garden up a long gentle slope that eventually levelled off into mile after mile of rolling, grassy downs. At the edge of this plateau the ground suddenly fell almost sheer to the road below, which ribboned away to Westbury in the distance. The horse had been fashioned centuries before and was supposed to commemorate some victory in battle. The turf had been cut from the steep hillside, to leave the prancing figure of a noble steed, picked out in the shining white chalk undersoil. The only turf left within the figure was a circle of green sward which was his unwinking eye. Some notion of the scale of the beast was that this grassy eye, which looked tiny but in proportion, from below, was six or more feet across, and it was possible to slither and slide along the chalky head and neck to take refuge on the ogling island of turf. Nothing delighted my child's mind more than clambering across with a packet of sandwiches and buns and doughnuts, and a bottle of old-fashioned burpy pop. Every year, from about the age of five to ten, I made my annual pilgrimage to picnic there but the fun wasn't finished even when I'd eaten and drunk my fill. I used to carry up a battered old tin tray to sit upon and slither down the summer slopes, faster it seemed to me than the helmeted Olympians go down the Cresta Run on ice.

Although both were family doctors, my uncle and my father had vastly different life styles. Dad was called out on Saturday nights to sew up the broken heads of colliers who had been brawling in Black Country pubs. He always dosed them with purgatives explosive enough to clear their bellies as well as their heads, and to keep them glued to closet seats so that, at least, he was never troubled twice by the same patients in one weekend. Uncle Tom's patients were far gentler countryfolk. Old John, his gardener, fashioned my first bow from yew, and arrows from thin hazel wands. Ted, a local shepherd, called regularly at the surgery, attached to the house, for physic for his wife – but never had a day's illness himself. His flock of sheep were fitted with sonorous bells, and he often took me up onto the downs, above the White Horse, for a whole day at a time, as he folded them onto fresh pastures across the chalky grass. He carried a bottle of sweet cold tea in his pocket, and cheese sandwiches wrapped in newspaper. Midday 'snappin' time', sharing Ted's tea and cheese sandwiches, was even more exciting than picnics on the White Horse's eye.

There were lizards basking on almost any fallen log – while we had only newts at home – and huge snails, with mammoth shells that thrived on the chalk-grown vegetation. The locals called them 'wall fish' and ate them avidly. The flowers on the grassy downs included acres of poppy and carpets of purple scabious, with scores of butterflies, quite strange after the limited variety of the Midlands. If we were there in September, I used to wander for miles across the open, unfenced downs, picking mushrooms for the family breakfast. Ted the shepherd shared his secrets with me of the best places to try.

We were usually there for the annual Westbury Sheep Fair – which wasn't really at Westbury but on the rolling downs above the White Horse, almost within spitting distance of Uncle Tom's garden below. Several thousand sheep converged from many miles around and the noise of their bleating was as continuous as waves breaking on the sea shore. Shepherds' voices floated faintly on the breeze as they greeted each other, after a year apart, or were sometimes raised in anger when strange flocks intermingled. The rural symphony was orchestrated by melodious sheep bells and by the staccato barks of sheep dogs to produce a medley of sound that is unforgettable. I used to wander up to the downs at half-past four or five in the morning, the day of the fair, to marvel at the kaleidoscope of shifting patterns as flock after flock arrived to be marshalled into the rectangular pens of wattle hurdles. They had been erected the week before with geometric precision and were removed for storage as soon as the fair was done. Such primitive barter needed no trappings, so there were no cattle trucks or megaphones or Musak or trade stands, which seem so indispensable to modern, high-powered cattle auctions. I didn't appreciate it at the time, but that ancient sheep fair can have changed but little since biblical times.

My uncle's house was a period piece of a different kind. It had none of the smart and fashionable features that are now the status symbols of prosperity. Nothing was chromium-plated, there were no cocktail cabinets, nor modern pictures, nor labour-saving devices. It was a model of shabby gentility, with comfortable but ageing chairs, and the pristine brilliance of the paintwork had long since been mellowed by tobacco smoke. After meals, washing-up meant simply what it said, and it was done by Milly, whose husband Frank was jack-of-all-trades, cleaning the boots, washing and servicing – and sometimes driving – the car and working with Old John in the garden. Water did not issue under pressure from the tap. It flowed, by gravity, from a tank in the roof which was replenished each morning from a well, the water being pumped by a hand pump in the kitchen. This was one of the jobs for young visitors, and I enjoyed it immensely for a week or so, though the novelty might soon have worn off if I'd had to do it at home.

The garden was also quite different from ours. It was a sort of picture-postcard garden, filled with multi-coloured flowers and butterflies, with crazy paving paths of local stone and white fantail pigeons to add movement and tranquillity with their soothing love songs. But it is the scent that I remember most vividly. The whole of one side of the garden was bounded by a purple lavender hedge that filled the evening air with almost tangible fragrance. There is a time of magic stillness on summer evenings when warm air ceases to rise as the temperature drops and chills it. The ripples on the garden pool died away for just a few minutes, so that the surface was as still as a reflective mirror and the sweet smell of lavender was almost overpowering. The grass suddenly sparkled with tiny lights that turned out not to be incandescent dewdrops, as they seemed, but love lights in the tails of glowworms.

As I grew older I could tramp for miles alone across the

velvet turf or along chalky tracks see-sawing over the rolling downland. Time and distance took on the illusion of infinity. The end of the journey promised to be at the summit of every rounded hill – but just when I seemed to achieve it, I always found, as in later life, that there was yet another hill to climb beyond the temporary horizon. The area was much used by the army for manoeuvres, and tanks came rumbling by as close and often as modern tractors would today. The village of Imber, a few miles from Bratton, had been evacuated and sacrificed for troops to practise the arts of modern urban war, pitching grenades through windows and fighting hand-to-hand from house to house. Yet, only a few miles across the downs, Stonehenge stood timeless as it had always stood. Alas, it stands like that no more today, for it has had to be fenced-off from mindless yobbos who care nothing for antiquity but are hell-bent on defiling the ancient stones with obscenities from aerosols.

The downs have changed as well. Velvety turf has given way to corn that merges with the skyline. The barley-barons, who pile up millions from their thousands of acres of corn, have banished the sheep to wilder, more rugged hills where tractors and ploughs would plummet out of control. Cultivation on such a scale would have been impossible up on the Bratton Downs when I was young but the development of earthmoving equipment and mammoth machines during the war sired a generation of tractors and implements that made pre-war ploughs seem puny. Huge multifurrow ploughs now surge across the downs, burying the ancient turf and exploiting its accumulated fertility.

Weeds are simply wild flowers in the wrong place and what is the right or wrong place for wild flowers is a subjective decision that depends on point of view. The barley barons reckon nothing to red mists of wild poppies or a blue haze of scabious, and spray the corn with chemicals to keep it clean. Profit is their yardstick.

So now the view is very different from the rolling sward of my youth. In spring, when the young corn shoots, it is greener than green, a glossy ad. exaggeration of our green and pleasant land. As the corn develops it is soused in more nitrates that turn the brilliance to a dowdy grey, which fades and yellows into what looks like an endless desert. If you don't mind monocultures – or misers' carpets of gold – you should own such a prairie that will fill your banker's vaults. And, when the corn is harvested, so much is being grown that there's no market for large quantities of straw that result, so it is fashionable to set fire to it and black out the changing countryside beneath a pall of smoke, though public anger is at last forcing even Big Business to mend its manners.

Those who can remember mile after mile of fertile downland, with snowy sheep and carpets of wild flowers, do not share the view that the good old days were really bad, or modern ways more civilised.

4

Flat as a Pancake

There is little visually attractive about fen country, which is flat and featureless, with belly-griping winds that rip mercilessly across from the east, unchecked since they left the Marxist steppes. Instead of hedgerows to divide the fields, there is a watery network of channels, which do not even bear the courtesy title 'dykes'. Literal-minded locals call them simply 'drains'. This, in fact, is precisely what they are – and therein lies the first charm about the place. This is a countryside as truly man-made as any Capability Brown landscape, laid out for the delight of aristocrats. Fanatics who scream for an unspoilt countryside would flounder in stagnant bog to their navels here, if it was not for the early civil engineers, immigrants from Holland, where such works were commonplace. These brilliant men accomplished with picks and shovels, centuries ago, what we should be hard put to do with bulldozers today. Before they came, this whole landscape was waterlogged and those inhabitants who did not die prematurely from fever or pneumonia or dysentery, seized up solid in their prime with the rheumatics.

The charm of fen country, as the charm of witty women, depends more on accomplishments than vital statistics. I learned to appreciate the fens on visits to an uncle who had recently retired as a bank manager in Spalding. It was in the 1930s, before farmers were feather-bedded, and one of a bank manager's perks was freedom to roam the land of farmers whose heads had been kept above water by financial help in the merciless slump. The black soil was as rich with fertility as the filter bed of a sewage farm, yet times were so hard that one of my most vivid memories was a standing group of corn stacks which had never been threshed because there was no market for the corn. (Shades of our 'intervention' charity!)

The custom in those parts was to build corn stacks in the corner of the field where it had grown and been harvested, instead of carting it to the homestead to be stored in a stackyard. This group of stacks had stood for four years where they had grown, so they had been colonised by rats, which had thrived and multiplied until they had literally eaten themselves out of house and home. When I walked into the field the whole ground skithered and slid like a carpet, drawn over a ballroom floor, as the army of rats retreated. They had worn all vegetation down to the polished black soil as they cut a spider's web of tracks out from the stacks in search of fresh food. My hobby had been ratting, since the Old Squire's day, but I would have thought twice about matching a terrier of mine against such an army. I was told that the farmer eventually surrounded the area with small mesh wire netting and set fire to the stacks. The rats could be heard a quarter of a mile away in their frenzy to escape their funeral pyre. It drove home the lesson to me that you can't beat the law of supply and demand, for those molehills of surplus corn, fifty years ago, were as unsaleable then as our corn mountains are today.

Judged by the standards of our Midland farms, the East Anglian farmers of those days were big and they either prospered or went broke in a big way, depending on their ability and luck. But there weren't any absentee farmers, such as the pension funds and finance companies who 'farm' huge areas today from impersonal offices in London or the Middle East. Those friends of my uncle lived on their land, either as owners or tenants, and they knew every square yard in busy times and slack, good times and bad. There was a wonderful rural community, where everybody knew everyone else, from the squire to the humblest farm labourer. It was an 'us' and 'them' world so that strangers were always under suspicion until a local vouched for them.

I was riding a very noisy racing motorbike in those days, usually illegally fast, so that it wasn't long before I was stopped by P.C. Dukes, the local bobby. His best friend wouldn't have called him quick-witted and he accused me of having an inefficient silencer, a charge he proved by pushing a stick up it to show it had no baffles. He took full particulars and had written an essay in his notebook before it emerged that I was staying with my uncle, who was the local magistrate. His whole attitude instantly changed, for he had been treating me as an already convicted criminal. 'I am sorry, sir,' he said. 'I hadn't realised where you were staying. Would you mind having it fixed next time you're in town?' So we repaired to the local and sealed the bargain over a couple of pints in the back room, where his uniform was less conspicuous. Real country places are still like that. Great store is set on being good neighbours – and it is not done to do a neighbour a bad turn. Larger communities are more impersonal.

Like many keen shooting men, my uncle put all his energy into outwitting and shooting his quarry. He did not spend the time necessary to have a well-trained dog. His spaniel was wilder than the maddest hare, and hopelessly disobedient. Once loose in the field, he applied the whole of his superb talents to flushing every species of game for his master to shoot. Unfortunately, minor matters such as the lethal range of guns were quite beyond his ken and he flushed every pheasant and hare in the parish long before his owner was anywhere in sight. My dog, on the other hand, was biddable. He did what was asked of him because

I don't suffer fools gladly. I am far more interested in having a clever working dog than I am in shooting game. So we made a perfect team. I was the unpaid beater, and my dog found skulking pheasants and rabbits or hares in forms but did not disturb them till uncle was ready and able to cut short their career. I didn't shoot because my marksmanship is such that I couldn't hit a bull up an entry, so I concentrated on my dog, using previous experience to predict the probable flight line of our quarry.

Although the land was so wonderfully fertile, the farms were very quiet. Horses had not yet been entirely replaced by tractors, which compact the alluvial soil far more than the hooves of horses did, and the two of us – and dog! – spent long quiet days in pursuit of game or simply watching it.

Although the fertile fens are very flat, Lincolnshire is a wonderfully varied county, extremely well wooded in parts. A friend of the family had bought Hagnaby Priory which included a lovely forty-acre wood which harboured long-eared owls and red squirrels, neither of which I ever saw at home. I spent long days mooching in the wood, alternately sitting, motionless, on some fallen log or clapped, statuesque, against a tree trunk, simply waiting and watching. Solitude in such secluded countryside is a priceless gift. Even the intimate chatter of the closest friends can send shy creatures scattering and, as I merged with my surroundings and became invisible, my mind and memory were filled with sights and sounds that have moulded my subsequent life.

That is half the fun of messing about in strange territory that harbours creatures which are not found at home. I remember Lincolnshire for sport on its open fenland and for the sublime peace of its quiet woodland. I remember it for its cohorts of rats, not only in the verminous corn stacks, but for colonies that made warrens in the banks of the dykes, the 'drain-rats' that emerged at night to forage on sugar beet and corn that grew in adjacent fields. The banks of the drains were steep and thickly covered with a mat of vegetation which was the perfect cover for pheasants, so we would walk each side of the drain, keeping my dog downwind where he could catch the faintest aroma of his game. We disturbed short-eared owls as well as pheasants, for rats are among their favourite quarry, and they skimmed along the field boundaries at dusk, helping the farmers by pouncing on young rats.

I often watched them hunting because I would return for an hour or so of 'eel-bapping'. This was a specialist sport I loved. I hate catching eels on a fishing hook and worm, because they swallow the hook and writhe round the line so that disentanglement is both messy and cruel. Bapping does not involve a hook. A number of large earthworms are bunched with a thick woollen thread and dangled in the water on a string attached to a stick. When an eel sees – or smells – them, it grabs a large worm and its teeth become entangled with the wool. A sharp snatch on the string – and the wily eel lets go. But, if the fisherman can control his excitement so that he gently lifts his bunch of worms smoothly out of the water, the eel will not discover his plight till too late. He will hang grimly onto his booty as he is swung gently clear of the water so that, by the time he discovers what has happened, he will be above dry land. When he lets go, it will be too late and he will drop into a pre-positioned bucket or traditional, open, upturned umbrella, from where he may be harvested with none of the tiresome trauma of disgorging a hook and unravelling him from yards of fishing line.

Such simple and rewarding sport is the stuff the countryside is made of. A few miles to the south, the shallow Wash is alive with waterfowl – and where there are waterfowl, there are both sportsmen and naturalists, whose interests often conflict. My pleasure in the countryside is inversely proportional to the number of people sharing it with me.

I detest crowds but I like people individually, provided that we have some interests in common. The idea of a day at Blackpool or Brighton, picking my way to the sea over recumbent corpses of fat women, sunbathing like sows in a wallow, brings me out in goosepimples.

However beautiful it may be from afar, too much of the coastline is geared solely to commerce for my simple taste. The Wash is not like that. It is so flat that water, draining from the land, enters by way of a network of tortuous channels, zig-zagging their way to the horizon. At low tide, the sea is almost out of sight and one can pick a way along the banks of drainage gullies out towards the sea. Although this may be possible, it is extremely unwise for strangers because, when the tide turns, it comes in far faster than a man can walk. The first indication of this is that water in the gullies gets deeper and the gullies wider. Channels that could be taken in the stride when following the tide out suddenly become unjumpable. Before a stranger knows it, he can be marooned a mile or more from shore with water creeping implacably to cut him off for good. So strangers don't venture out alone onto the saltings if they are wise. The attractions to do so are the wildfowl and seals that haunt the sandbanks far out at the tideline.

Some of the local inhabitants make a respectable living by selling their specialised knowledge. They hire themselves out as guides to bird watchers, photographers or wild fowlers. Knowing the topography of the gullies as intimately as the charms of their wives, they offer to conduct clients out towards the tideline far enough to shoot curlew or wild geese with camera or gun, according to their tastes. I had heard a great deal about these men, especially Kensie Thorpe, who billed himself as 'the goose man'. The standard practice was to take a spade and creep towards the tideline, stopping at a spot where regular flightlines were known to intercept. The soil was soft mud, so that it was easy enough to shovel sufficient into a channel to dam it from the land. A few more shovelfuls, about six feet away, would also dam it on the seaward side, to hold back the incoming tide till the downstream channel filled. The space between the two dams was large enough to hold two men, like soldiers sheltering in a foxhole, and deep enough to conceal them up to their shoulders. A few sprigs of wiry weed placed round the dug-out edge gave perfect camouflage from birds flying overhead. Literally thousands of duck and geese, redshanks and curlew and whimbrels settle on the sandbanks at the water edge and, when the tide turns and raises the level, they fly to banks still standing proud nearer the shore. If the guide had done his job with skill, many of them would fly over his sunken hide so that his clients could snap or bag them.

I had heard that Kensie Thorpe, the goose man, went one better than this. He followed standard practice by taking his client out across the mudflats and digging a foxhole when he arrived. But, instead of waiting for birds to fly over by chance, he encouraged Mahomet to come to his mountain by mimicking the cries of wild geese and curlew so accurately that they changed course and approached him to investigate. I had marvellous times with him out on the saltings of the Wash, with neither gun nor camera. We crouched in the foxhole till the rising tide shifted a group of birds, which often flew parallel to the tideline, instead of inshore, towards us. The direction made no odds to Kensie, whose musical cries were indistinguishable from theirs, even to the birds themselves. Time and again they altered course, winging their way over us, several times alighting to investigate within twenty or thirty yards of our hide.

Most exciting of all, to me, was to get within hearing distance, as well as view, of wild seals, surely among the shyest and most persecuted of our wild beasts. Once, on a very still day, with the faintest air eddies swirling from the sea, we got within smelling distance too. A party of seals had been basking a hundred yards or so away and, as the tide rose, they shifted over to a steep sandbank only a few yards away. We almost subsided into the squelching mud in our efforts to remain unseen till, lying still as *rigor mortis*, we could hear their indelicate belches and smell their fishy breath. Of such stuff is the countryside made. Pretty views and wide-open spaces would be insipid stuff without chaps like Kensie to breathe life into their rural charms.

OPPOSITE Hilton Park, Staffordshire. Before the M6 was built it was part of The Squires' estate. Now thousands of motorists only associate the name Hilton Park with a service station. But many of the little woods and coverts, planted to hold pheasants and foxes and to improve the view, are still there.

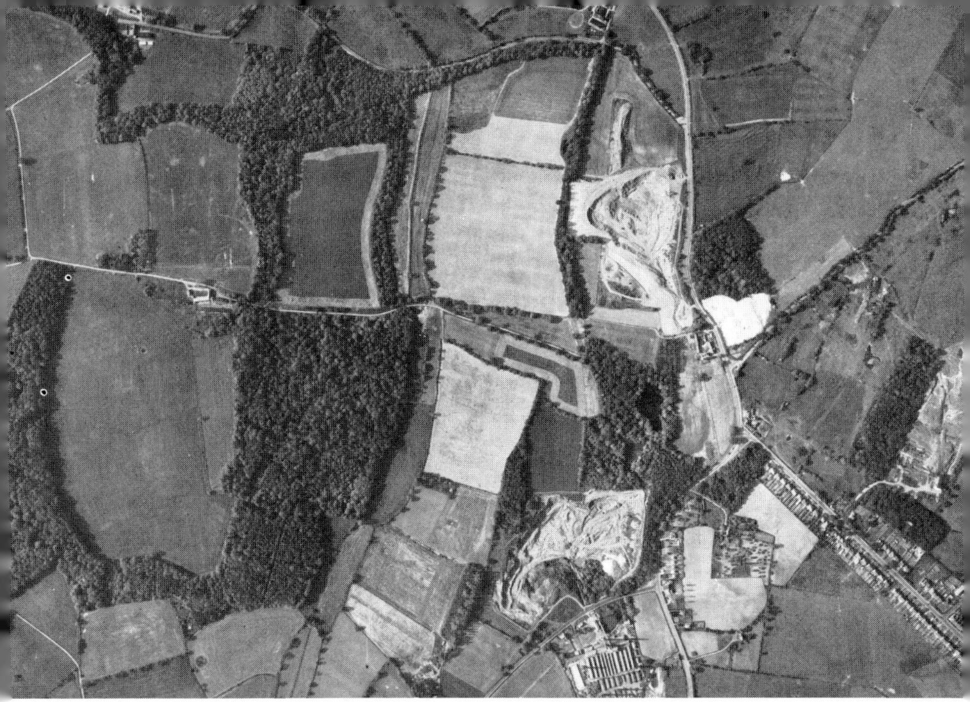

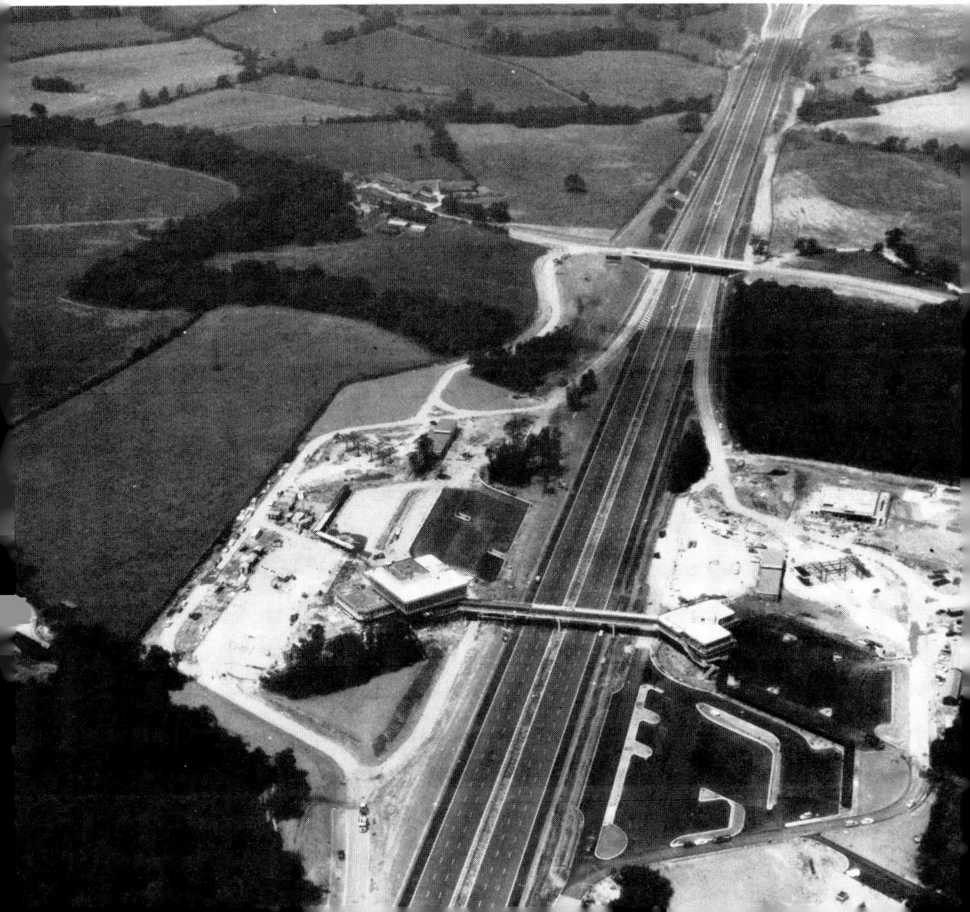

ABOVE The Westbury white horse, cut into the chalk downlands; as a lad Phil Drabble picnicked on the island of turf that makes the 'eye'.

RIGHT Modern hedge-trimming – most young trees do not survive the 'efficient' blade.

OPPOSITE, INSET Wicken Fen, Cambridgeshire, is one of the few remaining protected natural wetlands. Elsewhere in Cambridgeshire, on the fen edge, (MAIN PICTURE) modern farming methods dictate the need for prairie fields stretching to the horizon.

ABOVE The Brecon Beacons make a glorious backdrop to the patchwork of traditional fields and hedgerows in the Usk Valley in Wales.

LEFT The buzzard is one of Britain's most graceful birds, but is threatened by many sheep farmers.

5

Space

I know of no wilder winter country than the land around
Flagg Moor on the Buxton to Ashbourne road. If there is
a skither of snow forty miles to the south, where I live, it
is a fair bet that the main road up there is deep in snow-
drifts because it is usually one of the first roads in England
to be blocked and last to be cleared. In spring the air is
seductive with the lovesongs of curlew and golden plover,
the wildest but most tuneful songs in Nature. Wheatears
and whinchats, birds of open hill country, nest there in
summer, dipper-haunted streams gurgle over the rocky
beds of valleys, and precarious stone walls, more useful for
sheltering sheep than containing them, give the delusion
of eternity.

Urban reaction to the pressure and stench and frus-
trations of city life is the instinctive nostalgia to escape to
such deserted solitudes. In such hill country, wide skies
stretch from horizon to horizon, for the only trees to break
the view are scraggy clumps, planted in the vain hope of
giving a modicum of protection to the farmsteads. However
inviting such a wilderness may look, from the cosseted
comfort of soft city seats or heated motorcars, men who
scratch a living from such hostile soil know that the prediction of peace is an illusion. The farms are small and uneconomic because the rocky soil is so hostile that ploughing
is impossible. All that can survive are a few hill sheep
and cattle. Most farmers are part-timers, ekeing out an

existence by taking richer men's cattle to market, working for the council on the roads or finding sites for optimistic caravanners under the delusion that grass that side of the wall will be greener.

I love this country because it is so harsh that there is rarely much danger of being trodden underfoot by trippers, ruthlessly determined to be hearty and happy. The only day they will be there is for the High Peak Harriers point-to-point on Easter Tuesday. It is usually so bitterly cold that only hard liners stay for more than one of the five races, so sensible folk avoid it.

I found more fun when the same dedicated horsey folk met up there, in years gone by, for meets of Eric Furness's bloodhounds. Riding horses over such stone walls seems a masochistic form of suicide that never has attracted me. I prefer to be a sadistic spectator. Hounds met at the Bull i'thorn Inn or at an isolated farmstead and followers and spectators were invited in for rum and coffee while an athletic enthusiast laid the trail simply by running five or six miles cross-country, leaving scent to seep through his sweaty feet on the grass, through the soles of leather shoes that breathed. He drove on relentlessly, splashing through streams and vaulting stone walls in a fashion that would have shown up modern joggers as the effete eccentrics they are. By the time he got back to quench his thirst we had swigged most of the coffee and all of the rum he needed to revive him.

His vest was stripped off and waved in front of hounds as soon as he had safely disappeared, and they cast in a circle to find the same pungent stench clinging to the grass, inciting them to streak away with a cry as deep and sonorous as a peal of cathedral bells. Then the fun started for spectators and, I suppose, for the huntsmen, if they really liked their fun the hard way.

Horses are creatures of low IQ, who are obedient to Man only because they are directed by steel bits and prodded in the belly by the spurs of their riders. Even the winners of races get thrashed by their jockeys, however hard they try, though TV commentators take it for granted, with the casual remark that, 'the winner is now coming under pressure'. Intelligent dogs, on the other hand, do their masters' bidding, with no need for chastisement, simply because they enjoy being praised by the boss. The followers of the bloodhounds, 'putting pressure' on their horses to leap headlong over that unrelenting stone-wall country, were obviously a pretty unimaginative lot. Even when they came off, crashing on the crags, it didn't seem to hurt them much. It did not, of course, hurt me at all. For me it was a different form of escapism. I did not regard the view through the rose-tinted eye glasses of urban summer visitors. My mind tracked back to prehistoric men, who earned their keep by hunting long before their gentler descendants became farmers and domesticated their quarry. Now they confine their stock between stone walls, where it falls to butchers' knives instead of hounds. The banshee cries of modern huntsmen, cheering on their bloodhounds, echoed for me down the centuries so that the solace I derived from that harsh, deserted landscape was a satisfying sense of solitude, so rare in times of crowded insecurity.

Not all high country is so unsympathetic. The drive over the Eppynt, from Brecon to Garth, in central Wales, would make the most unimaginative feel on the top of the world. Mile after mile of rounded hills, the summits never higher than the road, it seems, roll on to the horizon. The army uses much of the land as a firing range and, however trendy it may be to vilify such use, it has advantages beside the obvious contribution to our safety. Occasionally red flags do fly, to signify danger not left-wing labour rallies, and courteous soldiers direct motorists to routes that are perfectly safe. But most of the time free passage is unrestricted, the views are – literally – out of this world and the solace is totally refreshing.

The soil is sparse and the grazing wiry, so that the pressure of economics would probably blanket the whole area under forestry conifers if the army did not want it. As it is, the land is untamed and buzzards glide effortlessly overhead. They are the most graceful birds who mastered the secrets of aerodynamic currents before our forefathers had crept out of their caves. With outstretched wings they soar ever higher on the thermals, rising or falling hundreds of feet without contracting a muscle. Sometimes a lone bird will swing high overhead, in arcs and circles, scanning the ground below for quarry that ranges from mice to rabbits. Sometimes whole family parties play rhythmic games of graceful aerial 'tick', calling each other with pleasant mewing tones that match the whispering winds.

Buzzards are among my favourite birds but, unluckily for them, they have hooked bills – and hooked bills or canine teeth are death warrants in the eyes of many stupid men. They equate them with danger to livestock or game, believing that creatures with canine teeth, from weasels to badgers, will attack and kill prey that should provide their own food or sport. By the same token, they believe that any bird with a hooked bill, whether hawk, falcon or owl, poses peril for their prized possessions, domestic or wild.

Like so many country beliefs, there is enough half-truth in the theory to give it credibility. Only fools would deny that foxes – and sometimes badgers – will eat domestic hens if they are easily available. Wise men shut their poultry pens at night so that their cherished birds do not come to any harm. The badgers and foxes, denied their easy, but unnatural meal, revert to killing rabbits and rats that are capable of doing so much more harm. On balance they are beneficial, but it is difficult to convince dog-and-stick farmers of the hills that foxes do not do them much harm because they have been reared on the dogma that all dead lambs were killed by foxes. The truth is that some are. When a ewe has twins, one strong and one delicate, the runt cannot compete for milk with his stronger brother, so the strong grow stronger and the weak, weaker. There comes a time when the weakling cannot keep up when the ewe takes her pride to richer pasture – and the foxes have a field day. If they had waited till the weakling died and tidied up the pasture by scavenging the garbage, they would still have been accused of ovine infanticide because so many countrymen cannot believe good of a fox. The fact is that such scavengers are needed to prevent casualties putrefying and spreading disease and animals which eat stillborn lambs and casualties perform a valuable service. Anyone who has seen a ewe defending her lamb against a savage dog will appreciate that foxes do not get many sound and healthy lambs. But hill husbandry is so slack that farmers are prepared to write-off a high percentage of their lamb crop because they reckon it is not 'cost-effective' to take better care. They reckon it all right to rear one lamb per ewe, while lowland farmers, who take the trouble to foster the second or third lamb from one ewe on to a ewe that has none or one, regularly rear an average of almost two per ewe.

The method hill farmers use to deal with the foxes that feed on their dead lambs is very simple. They apply to the Ministry of Agriculture for a permit to buy strychnine for moles – and use it to 'dope' casualty lambs which are put out as bait for foxes. Not only does it kill foxes, but dogs and cats and crows and ravens and magpies and any other species that are partial to young lamb for dinner. Those creatures that don't like lamb are offered rabbit, laced with the same poison, by courtesy of the Ministry of Agriculture, Fisheries and Food, which is the only authority legally entitled to issue permits for strychnine, one of the most persistent and deadly poisons spewed over the countryside, ostensibly for moles.

So the buzzards that I love on the wild hills in Wales are inversely proportional to the number of sheep there. Many hawks, including buzzards, are lazy birds which will eat the food that they can get most easily. They catch rabbits as their main diet when available, and rats and mice by hovering like kestrels and pouncing on them. But they are also scavengers and feed on carrion if they find it lying in their territory. Since there is most carrion, both in the shape of dead lambs and the afterbirth of live ones, in 'good' sheep country – carrion liable to be laced with poison – buzzards are often scarce where lambs are plentiful.

There are sheep on Army firing ranges which are supposed to be moved before major exercises, but many of them are poor old screws put there deliberately, because if they are accidentally shot the farmer is compensated. And compensation for scrub stock is nice business – if you can get it. So casualty sheep are not left lying around here but collected as evidence. Nor are lamb carcases baited with strychnine, for the laying of poison – which is illegal – would be asking for trouble on land frequented by soldiers. So the high hills between Brecon and Garth, besides being exceptionally beautiful and beautifully lonely, have solitudes shared by delightful, mewing, soaring buzzards.

It is hard to get to know country far from home more than superficially without staying in the area and pleasant pubs are part of the countryside worth cherishing. We used to stay at The Lake, a fishing hotel at Llangammarch Wells, about a mile from Garth. Just as the Eppynt is among my favourite hill country, that hotel was my idea of what real country hotels should be. No soft Musak nor bridge parties nor ponced-up clothes. Although it had once been a thriving spa hotel where our grandmothers went to drink the foul spa waters, it was clapped-out by the time I knew it. There were often more sitting-rooms than clients so that if any guests had nothing in common, there were always retreats where they could be avoided. The proprietor was apparently satisfied with a very modest living so that he often failed to put in an appearance for quite long periods. Some of the guests never saw him at all. We reckoned that the whole enterprise would have folded up altogether if it had not been for a Canadian Air Force officer who had been there – as a guest – since the end of the Great War. He was terribly scarred from burns, when his plane had been shot down, so that he couldn't stand the cold of his native land and had settled at The Lake, where he enjoyed the fishing. When the place was in decline, he had come to the rescue by running the bar and allocating beats on the Irfon, which ran through the grounds, for fishermen. He was obviously more interested in living a civilised life than in making mighty profits – and his idea of a civilised life did not include the company of guests he didn't like. He didn't suffer fools, either gladly or in any other way; nor braggarts, nor social climbers, nor snobs, nor know-it-alls – and he was not shy about telling them so in front of fellow guests. As he ran the bar, their drinks were a long time coming, and as he said where they should fish, they didn't catch many. As a result, the place was never overcrowded and, if he liked you, the chances were that you would like your fellow guests, because they were all he would tolerate.

I am no fisherman, but there were badgers in the grounds and pied flycatchers and red squirrels. A pair of buzzards nested in an oak above the hotel and dippers in a crevice of the decaying spa. It was a marvellous place for naturalists to stay and my wife and I used to take our whippet Dinah and catch rabbits on the hills above, as we went for after-dinner walks. We went mushrooming before breakfast and badger-watching when dusk fell. The last time we went, the Canadian was dead and the place had been taken over and smartened-up, with many more guests, some of whom would not have lasted long in the good old days. But there was a better choice of food, bedrooms with a bath, and a miniature golf course and tennis courts to siphon off the surplus.

I always liked the story – probably apocryphal – of how the place was started. A brilliant up-and-coming young London financier got the tip-off that the City of Birmingham was about to buy up land in Wales to build a dam to supply the city with water. The land, at the time, was worth virtually nothing – but it would be worth a bomb when the scheme was finalised. It was all said to be in the valley of the River Irfon. So the young financier was despatched to Wales to get his hands on as much land as possible from which a killing could be made when it was resold. He succeeded in buying up a number of farms until the last item on his list was The Lake Hotel on the river by the spa, though it fetched more than he would have wished. Only when he had completed his assignment did he discover that the original tip-off had been inaccurate. The Birmingham waterworks were to be in the Elan Valley, not the Irfon Valley. The young financier had bought the wrong valley, and as a punishment for his sin he was banished from his office in the Big City, to languish as manager of a clapped-out hotel at the back of beyond. I have no idea how much – if anything – there is in the story, but the proprietor, who was not often seen, was very smart and dapper. He looked the part of a financier! And vast tracts of countryside *do* change overnight at the whim of city gents who are puppets of the profit makers. Their blunders as well as their inspirations can change the whole aspect of the countryside.

6

National Parks

Few would quarrel with the concept of National Parks except, perhaps, the people who live in them. It is easy enough to be generous with other people's goods or land but, despite their title, National Parks are neither parks nor national. They are large tracts of open, upland country-side in England and Wales, but not in Scotland. Most of the land is in private ownership, farmed in exactly the same way as similar land in other parts of the country. It so happens that there is little ploughland because all parks designated so far are in the uplands where the harvest would be late and the high rainfall would prevent it ever drying if it did grow to maturity. Except in a very few areas, where access to open moorland or hillside has been *negotiated* with the owners, the access is limited to public Rights of Way, as in the rest of the country. Farmers and landowners, who find stock straying because gates have been left open or stone walls have been vandalised, resent trespassers exactly as most other people would do if they found strangers having a picnic on their front lawn and tossing their litter onto the flower beds.

William Wordsworth, who lived at Grasmere in the Lake District in the early 1800s, foresaw that the arrival of the railways would bring with it 'persons of pure taste' who would deem the district a sort of national property – 'and derive great pleasure from it'. But he also appreciated that the Industrial Revolution would inevitably mean more

43

leisure for travel by rail, so that 'the masses' would overrun his beloved Lakeland and destroy it by sheer numbers. What the railways began, as the shrewd old poet prophesied, the M6 motorway is in danger of completing. The narrow Lakeland stone-walled roads become yearly more constipated by the cars that cascade off the motorway.

The parks, as we now know them, were sired in 1945 by John Dower, who defined National Parks as '. . . an extensive area of beautiful and relatively wild country in which, for the nation's benefit . . . the characteristic landscape beauty shall be preserved and facilities for public open air enjoyment be provided.' He continued: 'Wildlife and buildings and places of historic interest shall be preserved.' Finally, 'Established farming use shall be effectively maintained.' Where Dower and I differ is that I should have work – and farmers' continuity – higher on my list than play or leisure.

The 1949 National Parks and Access to the Countryside Act proposed twelve National Parks, ten of which have been established. The first was Peak Park in April 1951 and the last was the Brecon Beacons, six years later. The others are Dartmoor, Exmoor, the Pembrokeshire Coast, Snowdonia, Yorkshire Dales, Lake District, North York Moors and Northumberland.

Wordsworth's gut reaction that the pressure of masses of humanity would stifle the very thing they loved was a depressingly accurate prediction. The proposal to designate a park in the Cambrian mountains was defeated by the intensity of opposition from local inhabitants, including farmers, who had seen what happened elsewhere and were determined not to be trampled to death by trippers. This was not an unreasonable fear because favourite areas of the Peak Park, for example, have been irreparably eroded by continuous wear by tens of thousands of human feet. Replacement of the natural surface by concrete paths makes a mockery of the whole idea of escape from the ratrace to peace in some refreshing natural wilderness that is as it was in prehistoric times – and so will forever remain.

Various remedies have been tried. The nostalgia for escape to lands our ancestors loved is so superficial that research has shown that a high proportion of visitors to quiet and lonely places never venture more than a few hundred yards from their cars. They produce deck chairs from the boot and the elders snooze and picnic, often leaving litter when they go, while others shatter the serenity by turning up the volume of pop music on their radios or damming streams with stones, torn from craftsmen-built boundary walls, or leaving open gates for stock to stray. The authorities have tried to contain this lunatic-fringe in the Peak Park to areas where they will wreak least havoc, by closing the Goyt Valley to motor traffic at weekends and Bank Holidays. Experience has shown that the most disruptive elements are also the most idle. They will go anywhere a car will carry them, using legs only to prop them up against the counter in bar-parlours, or to stand on while they vandalise stone walls.

To help those who can appreciate the solitude, the park authority runs mini-buses, which decant visitors who enjoy walking in the countryside at strategic spots from which to explore the pleasant places. But, every year, in dry spells, whole areas of the moor have to be barred from the public because the careless start fires by dropping matches or cigarettes and the evil ones do so deliberately. Until the government is prepared to get really tough and discipline such knaves and fools effectively, it is quite wrong to let them loose, however much the innocent, who are excluded, have to suffer for the guilty.

The signs are that the result of the present new industrial revolution will be an explosion that will turn leisure into the greatest growth industry of all time. Provided that we succeed in making the machine our slave and not our master – and there is little hope for our future if we don't

– the great problem will be to kill boredom by finding constructive and pleasant things to do in pleasant places. By the same token, it will grow harder and harder to earn a living doing the simple, labour-intensive tasks of hill farming that can't be mechanised. So why not combine the two? Why not use the leisure and tourist industries as a cash crop to help the men with small hill farms eke out a slender living from their land? One of the pipedreams of those who live in cities is to escape to enjoy the simple (?!) life in wild country, far away from their fellows. The same may be said of sportsmen, fired by the primitive urge to share with their ancestors the thrills of the chase. Shooting men and fishermen are prepared to lease sporting rights from farmers and landowners so that they may catch fish or shoot game as if they owned the land on which it lived. This is regarded by many farmers as a useful cash crop and those who scrape a living from hostile upland soil are more in need of subsidiary contribution than most. If they go out of business, the land they farm will either be afforested with softwood or revert to scrub or bracken.

So part of the price we are prepared to pay to keep the countryside we love might well be to dip our hands into our pockets to pay for the privilege of enjoying ourselves there, precisely as shooting and fishing men have done for generations. The added contribution might well tip the balance from loss to profit for those at present trying in vain to scratch a living there, so helping to preserve the landscape undefiled. The something-for-nothing school, who are forever whinging about their 'rights' over other people's land, should realise that enjoyment of the countryside is cheap at any price, and we tend to value most what is some sacrifice to buy.

Like so many pressures on the countryside, success in preserving beautiful places and encouraging folk to refresh their souls there can be very dangerous. The Peak National Park, for example, has seventeen million people living within fifty miles and half the population of England within day-trip distance. On fine Bank Holidays its beauty spots can literally be worn down to the subsoil.

Cars can be excluded from the most sensitive spots, as they are in Goyt Valley, but that simply spills the tide of humanity in other directions. One effective planners' ruse has been to create 'honeypots' somewhere else – or to grant planning permission for entrepreneurs to create rival attractions in areas that would normally be inviolate. This has been done a few miles away from the Peak Park at Alton, where the amusement park at Alton Towers busks-in a high percentage of those who never went a hundred yards from their cars in previously quiet beauty spots. Traffic jams six or nine miles long clog up the quiet country lanes, but ovine mentalities don't mind queueing. We have all had the thrill of finding an attractive quiet retreat shattered by some oaf who, seeing us settle, is convinced he must be missing something – he knows not what – but squeezes in beside us, just in case. So those who create honeypots to seduce the unimaginative to settle in swarms to be stupefied by mass entertainment are doing a public service – except to those who live near the commercial honeypot! They, poor souls, get caught in the queues, not for spoon-fed entertainment but simply to get home – or to get away from home. They are driven mad by the piped Musak that lulls the addicts into soporific stupor. The value of their property plummets through the bottom of the bucket. Their friends cannot get to see them nor they to see their friends. A small price to pay, perhaps, as a contribution to keeping other beautiful places undefiled – provided it is someone else who has to pay the price.

The sheer erosion, by the intolerable pressures of humanity, is not confined, of course, to the Peak National Park. More than half a million walkers tread Snowdon underfoot every year, a continuous, serpentine pilgrimage of a thousand or two foot sloggers, clambering and slithering to the summit and back on any busy summer day. A rack-and-pinion railway hauls another thousand to the sleazy summit, aptly described as 'the highest slum in Britain'. The National Park Authority is doing its best to tidy things up, but some measure of the problem can be seen from the fact that almost a quarter of a million pounds a year is spent to combat wear on the footpaths alone.

Northumberland National Park perhaps offers the best prospect for the escapists who really do want 'to get away from it all'. It is wild border country, not well served by motorways nor within too easy reach of densely populated industrial regions. Critics of the Forestry Commission will resent the fact that about a fifth of the park is blanketed in fir trees and those who dislike the army will be equally critical that they control about the same amount, which is used intermittently for firing ranges. But, in our crowded island trees have to be grown somewhere – as it is we import most of our timber and have a far smaller proportion of our land as forests than continental countries. And if the army is to protect our way of life, it has to train. When the firing ranges are not being used, which is most of the time, the public is free to walk there and the Keilder Forest is a shining example of the facilities for wildlife conservation – and observation! – and lovely waymarked walks the Commission so often provides.

In many ways Northumberland is a vivid reminder of a past that fuels escapist nostalgia. It is the most thinly populated area in England so that wandering here gives a vivid tang of the sort of landscape we instinctively pine for. There are still red deer in the woods, descendants of the deer our ancestors hunted, and the last remnants of our native wild white cattle still survive at Chillingham. My wife and I journeyed all the way to Northumberland specially for the thrill of seeing them and they are still so wild and undefiled by 'progress' that they are extremely dangerous and will attack ferociously to defend their young. The ramblers who moan at the prospect of an inoffensive Hereford bull in a field would have something to get in a twist about if they came face to face with the wild, white cattle of Chillingham!

The remains of the Roman wall, built by Hadrian along the Scottish border, is redolent of the distant past, in quiet places, too. The conflicts and disputes about our modern countryside are kids' stuff compared to those 'good old days' which put them in perspective. Exciting to relive in our imaginations – though certainly no fun at the time!

So what of the future of the National Parks? There is no shadow of doubt that the explosion in leisure and easy transport will impose ever greater pressure to escape to quiet places. But how can they be *kept* quiet and secluded, when so many people flock to them? Perhaps it may be possible to siphon off the surplus who doubt their own judgement if they arrive somewhere that is not crowded: the type who think village pubs are no good if they don't have to queue at the bar, wait an hour for a table and scream over the din of hi-fi to make themselves heard. *Homo sapiens* is so gregarious that a high proportion genuinely enjoy crowds, so it is sensible to provide diversions specifically targeted towards addicts of mass entertainment. But let us try to site such honeypots where they will not interfere with the peace of countryfolk, who have settled in the country to satisfy their craving for a quiet life in secluded places.

The advocates of National Parks frequently moan because the government, by which they mean the taxpayer, keeps them short of cash, about three-quarters of which is supplied by central government and the rest by local ratepayers. MAFF, the Ministry of Agriculture, Fisheries and Food, which does more to destroy the flora and fauna of the countryside by telling farmers which poisonous chemical sprays will kill 'pests' most effectively, and which gives grants for removing hedgerows and trees, and gets more taxpayers' hard-earned cash than the National Parks Commission does to preserve the countryside. MAFF stumps out huge sums to destroy the countryside by ploughing ancient grassland, denuding the countryside of wild flowers, and draining water meadows and wetlands to create ever bigger mountains of surplus grain and bogs of surplus butter but, if the money was channelled into National Parks, how would they spend it?

Bureaucrats being bureaucrats would, I fear, try to grow more and more important by attracting more and more people to their parks, thereby defeating the whole object. If they would spend the money on projects to persuade as many as possible to enjoy themselves on mass entertainment, it would take some of the pressure off the parks. The next objective could then be to persuade those who longed for space and solitude to spread themselves as thinly as possible over as wide an area as possible, so that they could find the isolation they craved.

The original Youth Hostel Association was a wise idea. A chain of hostels was set up across the land so that hostellers could walk or cycle from one to the next, enjoying the countryside all the more for the effort expended to see it. The hostels were self-service, even spartan by modern standards, so that the life they offered was simple, but the comradeship this engendered amongst those who used them was all the warmer and more friendly for that. The YHA is still going strong, of course, but its greatest admirers would concede that conditions are softer and more modern. But if the use of National Parks could be angled for those whose real need was for quiet and beautiful places, who were prepared to expend effort to make their pleasure worthwhile, as the Youth Hostels Association did, it might well be better for all. It would give a better chance of tasting refreshing solitude for those who use the parks, and it would cause less disruption to the lives of those who live and work there.

7

A Grant of Land

Twenty-odd years ago I was asked by BBC News to cover the sale of a large estate, which had belonged to the same family since the fourteenth century. It was described, in the auctioneer's catalogue, as 'the remaining portion' of the estate but even so it had about a thousand acres of woodland and almost another thousand acres of farmland, as well as a few cottages. The Big House and a handful of acres round it was not for sale as the last owner's widow still lived there. The assembly room at the village pub was packed for the sale, mostly by spectators because there were few prospective buyers. The whole place was pretty run down because the last few owners had not been direct heirs but cousins, who had had to pay heavy death duties and, perhaps, had not as great affection for their land as if they had been bred and born there. There is less incentive to leave an estate in good heart for cousins than for direct descendants.

Although the auctioneer did his best with his unenthusiastic audience, he eventually knocked the whole lot down for £85,000, which was the highest bid. The woodland was relatively worthless because a previous owner had sold ten thousand prime oaks, in the hungry 1930s, for a paltry £10,000. They would now fetch more than the whole estate realised. All that was left was the rubbish, not worth chopping down when the prime timber was sold, so it stood in rough and feggy parkland, as pathetic as the war-torn

51

fields of Flanders. The woodland itself had been let to the Forestry Commission on a lease of 999 years. The rent was half a crown an acre – twelve and a half new p, in the miserable washers which now pass as currency. It was not surprising therefore that land which had been squandered to bring in a paltry £125 per annum for a thousand acres should fetch next to nothing at public auction.

The farmland was knocked down for about £80 an acre to property dealers said to be based in the Bahamas, who paid the deposit, split up the land and resold it in lots, at a profit, without ever having to find more than ten per cent of the capital!

Being present in that homely country pub, on a golden September afternoon, to see an ancient estate disembowelled by strangers was like being witness at a public execution. It was very sad and very dramatic, and it set me wondering about the main causes for the decline of the great estates which have been directly responsible for so much of the pattern and personality of the countryside. I wondered just what impact the descendants of those who had large grants of land long centuries ago have made upon the countryside and its inhabitants. So I suggested to a producer on Steam Radio that we might take my recording of the actual bid that spelt finality for that estate and use it to focus the precise moment of disintegration, the final explosion that changed the face of that large slice of countryside as dramatically as any bomb could do.

Two farms were sold by the property dealers to sitting tenants; and the old Park, a farm and most of the woodland was sold to a turkey-tycoon. He uprooted the stag-headed oaks to pious cries of horror from the trendies. The fact that they stuck out like rotten teeth from a pock-marked face and were an eyesore to all but those with surrealist vision meant nothing to protesters. Nor did the fact that the park had degenerated through generations of neglect to scrub and feg and bracken that would not feed a mountain goat. Those who dislike modern farming do so on principle and are prepared to peddle their cause till logic flies out of the windows. But it is never safe to generalise. There are bad modern farmers – and there are good – and the tycoon, who was pachydermatous enough to ignore opposition, carried on. When he had grubbed out the scrub trees, he laid land drains over the whole area, ploughed it, limed it and sowed it in rectangular paddocks of a hundred acres apiece, divided by wire-mesh fencing. The worst type of prairie farm, you might think. But when he had done, he planted about 10,000 trees as shelter belts and game coverts. Not just miserable foreign pines, but mixed hardwoods with pines as a nurse crop to shelter them when young, and encourage them to grow into tall and beautiful trees. When the pines mature they will be sold as a crop, leaving the native hardwood to shelter and beautify the park. A flock of 1,000 sheep were folded by rotation round the farm to impart fertility to the previously sterile, feggy soil, so that within two decades the desolation of neglect has been transformed into a productive farm which offends no eyes except the incurably prejudiced.

The instant of the sale of that estate could be the focal point of a series. I suggested to my producer that it might be fruitful to examine the effect of those who had held other estates, by a grant of land, for generations; to assess what contribution their past had made and what the prospects were for generations yet to come. By dipping into records and talking to landowners and their servants and tenants it might be possible to construct a jigsaw that would depict what landowners had put into the countryside and what they had taken out.

The idea took off and we chose estates in different parts of the country which were in very different circumstances, keeping the wolves from their doors by very different means. One small Hereford estate, of only about 600 acres, was owned by a distinguished geographer, whose ancestors had been there for centuries before him. Like so many other estates, he said, 'the valley was not a famous valley. It did not produce famous people. No epoch making events took place in it, though it had its part in many. It did not provide any more lasting monuments than it still displays: the persistence of rural life over a thousand years, with the same recognisable structure and foundation it had before the Norman Conquest.' The countryside there had all the attractions that are missing from our modern insecurity. Beautiful, well-wooded *natural* countryside, immune because of its position in the basin of a valley from all the abominations of modern, mechanised chemical farming. It had the romance of the ownership of one family which proved that such continuity is still possible in our era of unpredictable change. The kind of loving ownership that does not march hand in hand with sordid commercial priority. There was the running water of pure streams and ageless hills above, good cattle country to delight the eyes of instinctive stockmen. If I, as a stranger, could fall in love with the place, how irresistible it must be to one whose sires and grandsires loved and tended the same land.

As a distinguished geographer, the owner, Lord Rennell of Rodd, decided to find out even more about his land than other people knew. He devoted ten years of his spare time to writing a book, *Valley on the March,* not only about what was there but, far more interesting, *why* it was there. Much of his information was the verbal lore of countrymen, handed from father to son by word of mouth; facts unfathomable from books; facts which are only known by those who have ploughed and sowed and reaped the land themselves. He discovered why each field is the size and shape it is, why hedges are so placed, and why the paths meander where they do.

It reminded me of the narrow tracks that snake, apparently aimlessly, across wild hill country, instead of spanning taut between two points, as tidy-minded planners scribe their artificial waymarks upon their impractical maps. Shepherds call these hill tracks 'sheep-trods', because they are worn by generations of sheep wandering from one pasture to the next. And sheep, being far less stupid than they are supposed to be, pick the *easiest,* not the shortest route between two points. So did the paths on this estate. Two hundred years ago the population of England was about a tenth of what it is today. So, in remote countryside, nobody minded neighbours using whatever ways were easiest to cross their land for trade or worship or to court some wench a mile or two away. In any case, the passing neighbours did not pass on without leaving behind the latest gossip, so that paths were not only physical communications but vital news media too.

It is quite unreasonable for strident pressure groups to claim that, because rights of way are 'ancient', strangers should be free to march, as of right, through farms and fields as friendly neighbours did. A few welcome neighbours should not set a precedent for hordes of interlopers in quite different circumstances. The ancient paths should now be rationalised to minimise the inconvenience they cause to those who live and work upon the land.

Patient detective work at this 'valley on the march', disclosed, from hedgerow trees and the banks which joined them, where field boundaries had been in the past, and it is fascinating to discover that they were *not* always as they are – as fanatics who decry the changing patterns of modern agriculture pretend. When horses replaced oxen for ploughing, they were more manoeuvrable and could work effectively in fields with shorter furrows. So hedges were grubbed up and field shapes changed to suit the modern methods, just as they are today, only this time the change is to cope with mammoth machines. I wonder if our forefathers regarded change as progress or desecration of the countryside?

Because this Herefordshire land was largely water

meadows some distant ancestor had decided to irrigate the land when he wanted, instead of as an act of God. When the owner took me round he showed me, with great pride, how he had restored a leet, or diversion of the river, which had once been cut along one side of the valley, so as to keep the flow of leet water above the valley bottom. He had restored sluices and channels, which were cut from the leet to the river on the far side of the valley so that, by closing one sluice and opening the opposite one, he could flood his fields for as long as he wanted. When they had absorbed the water they needed, he opened the bottom sluice and closed the top one, to drain off the surface water. He had an irrigation system as modern as tomorrow, not made by bulldozers and civil engineers with powerful pumps, but worked by gravity and the practical genius of his ancestors. That was the sort of countryside that is quite beyond price. It is beautiful and traditional, functional and steeped in history. But we saw other estates, that had originally been grants of land, that were very different.

One man, another noble lord, lived in a castle and was the ceremonial Black Rod in Parliament. He took us into a room over the castle keep where the tactics for the Battle of Edgehill were hatched. He told us of his ancestors' part in wars between Cavaliers and Roundheads and was more interested in what his ancestors had contributed to history than to agriculture or rural life. His was a marvellous castle and procedure in parliament might now be different – for better or worse! – if it was not for his illustrious forefathers. But his interests did not seem to lie in his lands and he never even showed us round his estate.

Another landowner had decided that making an estate a self-supporting unit was not commercially viable, so he spent five days a week in London, working as a financier, to make enough brass to keep his wife and her hunters at home on the estate, in the style to which she was very obviously accustomed.

A Grant of Land

Another family were here before the Normans came so that we were astonished to discover almost no continuity with either servants or tenants although it was a large estate of almost five thousand acres. The snag was that the family were Catholics and the landowner himself had joined the Church of England two generations ago. The result was explosive, farms changed hands and servants were sacked. More strife has been caused by religion than cured by faith. This estate is physically satisfying – apart from having a motorway driven through it – but a high proportion of those who earn their living from its land have their roots sunk deep in other parts.

None of the estates and families we examined for the series were in the Big Time League of Stately Home owners. We deliberately chose small fry, partly because the break-up sale that prompted the idea was obviously signposted 'Way Out'. 'Shirt-sleeves to shirt-sleeves, in three generations' is often true of industrial entrepreneurs. The first generation is fired with enough motivation and get-up-and-go to trample on the opposition and make a success. If not, nobody would have heard of them for only success makes news amongst such whizz-kids, the failures sink without trace. The second generation, brought up to work hard by a father who thinks there should be more than twenty-four hours in the day, maintains success or improves it. But, having been thrust above his station, the son is determined his sons shall have better chances – by which he means more leisure and more money. So the inevitable happens. The third generation leaves the sordid business side to underlings, who either slack or rob him till his business folds and he goes up the spout.

With landowners, things are only slightly different. Having survived in a highly competitive world myself, I have not much sympathy with those who start life with a handful of trump cards and let their treasures slip through their fingers. Too often, money is a dirty word to them.

They don't mind squandering it but would think it beneath their dignity to earn it. So they leave agents to run their estates, who deal with rents and tenancies, mining rights and repairs, and the whole commercial aspect of what is very big business. Such men grow very powerful and frequently change places with their employers in all but title. 'Lord Agent' and 'Mister Owner' is all too common and is the sure preamble to the decline of yet another great estate. This has a tremendous effect on the surrounding countryside, quite out of context with the success or failure of a single family, however ancient their lineage may be.

I happen to live on the edge of the ancient Forest of Needwood. My nearest shopping town is the pleasant market town of Uttoxeter, six miles away, with Rugeley on the edge of Cannock Chase six miles in the other direction. My nearest mainline railway stations are Stafford and Burton, each twelve miles away. It is in the middle of a huge slice of country about eighteen miles across which has remained undeveloped and unspoiled, not because of bureaucratic planners but because it was spanned, till recently, by a number of large, contiguous estates, where the landowners did not want them spoiled, nor did they need the cash to force their hands. The estates of the Earls of Shrewsbury, Harrowby, Lichfield and the Marquess of Anglesey joined lesser lands, held by minor landowners, to form a solid, impregnable jigsaw. None of them wished to see the lands of their fathers despoiled by jerry builders or big business, so they resisted development as fiercely and far more effectively than conservation-minded planners. It was, until recently, a magnificent, unspoiled tract of well-wooded countryside.

Since the last war, the landowners there have collapsed as helplessly as a house of playing cards. Death duties – double death duties in one generation, in some cases – spelt finality for some. Scandals, which split up families and channelled fortunes away in legal fees, tipped-up others who should have known better. Incompetent or dishonest agents and owners who believed they could suck a quart from their pint pot, were equally sure candidates for disaster. The long and the short of it was that estates here, as elsewhere, broke up, and those who bought them stripped them naked, to repay the purchase price, or fragmented them to try to make a profit.

8

Top People

Not all great estates, I am glad to say, have been emascu-
lated by taxation or greed, scandals or incompetence. Some
of the finest are still alive and very well. If I had my choice
of the most beautiful places in the world, I should not settle
for any foreign clime, the rugged grandeur of the Lake
District or solitudes of Scotland's Highlands. I would
choose, instead, some nobleman's estate. I wouldn't covet
his house, for the stately homes of England mean little to
me. They are anachronisms, relics of days gone by, when
labour and fuel were at rock-bottom prices. Draughty,
unheated barns, however lovely, do not turn me on. My
ideal would be a modest, labour-saving house, no bigger
than the one I own, but set in the heart of lands mellowed
and cared for by generations of the same family.

The Dukeries in Nottingham, or the estates of Longleat,
Woburn or any of a dozen more, all fill me with envy for
the minds of the men with the vision to choose such spots
to build their houses. The views and vistas are superb.
Woodland and water and parkland fill my eyes with pleasure. But put yourself in the place of the chap who first
stopped there and said to himself, 'This is the spot for me
and my heirs. We will live here evermore.' Did such men
start with empty landscape and plant the trees, in clumps
and larger woods, to build up a pattern that they visual-
ised? Or did they start with solid forest and nip and prune
spaces for their parks? Those who started with space really

were planting for their heirs, for they could never have lived long enough to enjoy the fruits of their imagination. Those who literally carved estates from primitive forest must have had the gift of second sight to be able to visualise what they would see when they chopped down great trees which blocked an even better view. Mistakes would be irreparable because trees felled in minutes take generations to replace.

Compared to such visionaries the fighting men who lived in moated castles, were born on Easy Street. Picking an impregnable height or unassailable rock to build a dour castle might satisfy the unimaginative mind of a mercenary soldier, but he would be welcome to his barracks when he'd built it, so far as I am concerned.

Among my favourite estates is Chatsworth, in Derbyshire, the home of the Duke and Duchess of Devonshire. I have an instinctive nostalgia for the area because, though I am a Staffordshireman by birth and preference, my grandfather owned quarries at Darley Dale, near Chatsworth, and I was brought up on stories about the estate. High above the house is Bess's Hunting Tower from which Bess of Hardwick could watch hounds hunting deer below when she grew too old and stiff to follow the chase on horseback. I sometimes think the tower answers part of my question about the mechanics of planning such estates because the view from the summit of the hill where the tower perches is so panoramic that, even with such well-wooded countryside, it might have provided a bird's eye perspective, clear enough to plan the future layout.

Like boys of all ages, I love fire and water – and the fountains and cascades at Chatsworth are beyond compare. The story is that an earlier Duke was visiting the Emperor of Russia, who showed him a fountain of prodigious height. The Duke was much impressed so, when he came home, he told Paxton (later Sir Joseph Paxton), his head gardener, he wanted a fountain to beat it. Paxton was a man of character. He didn't let the grass grow under his feet. When he came to Chatsworth, as Head Gardener, in 1826, he arrived at four-thirty in the morning on 9th May, climbed over the wall, explored the garden and set the men to work at six o'clock. Then he went to breakfast with Mrs Gregory, the housekeeper, and her niece Sarah Brown – to whom he took an instant shine. He married her the following year! So a minor matter like a fountain must have seemed pretty small beer, even when his grace added that the Emperor was coming on a visit in November to inspect it. There was no question of using pumps to power it, so Paxton started by building dykes and conduits to bring water for two and a half miles across the neighbouring moor. Immediately above the house, several hundred feet above it, he constructed the eight-acre Emperor Pool for final storage, with a chain of subsidiary lakes to act as back-up. The drop from the Emperor Pool to the formal garden by the house, where the Duke wanted the fountain, was several hundred feet, so that the water pressure generated would be tremendous. Calculations to decide the dimensions of the pipes were an exercise in advanced hydraulics.

I was so fascinated by the whole conception that I persuaded a television producer to do a programme about the problems involved, behind scenes, in running a great estate. Naturally, I bagged the section on the waterworks for myself. It boggled my mind. To design and construct the dykes across the moor and the chain of storage lakes, even with the latest earth-moving machinery, would make modern civil engineers scratch their heads. Paxton only had gangs of navvies, with picks and shovels, and horses and carts to shift the spoil. But militant, moronic shop stewards were as yet only plagues of the future. Craftsmen still took pride in the quality of their work, without whimpering about Them and Us and the possibility that Them might own more of life's goodies than Us. So Paxton and

his men worked in shifts by day and, with flares, in shifts
by night. Even under such pressure they found time to
stone the banks and to leave ornamental islands because
their pride was such that they wanted to create a legacy
for the future as well as use for the present.

While his men were digging and carting the spoil, Paxton
was doing his sums to calculate the size of pipes he needed,
and chasing up the finest iron masters in the land to manu-
facture them. The Emperor Pool, where water was marshalled from the moors, is several hundred feet above the
fountain, so Paxton ordered iron pipe, fifteen inches in
diameter and weighing 217 tons, with walls more than an
inch thick, to withstand the pressure. Paxton succeeded,
with cart and horse, muscle power and instinctive know-
how, in being ready by the time the Russian Emperor was
due to arrive. He really had something to show him because
his fountain was, and still is, the highest gravity-fed fountain in the world, being capable of throwing its jet two
hundred and ninety feet in the air. Alas, the best-laid
schemes may not succeed and, after all the work and hassle,
the Emperor did not turn up. Rumour has it that he got
as far as Yorkshire but, when some arch-stirrer suggested
that he would have his eye wiped, he did a U-turn and
scuttled back to his homeland.

The Emperor Fountain is but one of the watery marvels
at Chatsworth. The most spectacular is the Temple, where
even the domed roof is a circular waterfall, with jets below,
playing in rainbows to curtain the entrance, eventually
uniting to clothe the long series of steps, over which the
water cascades down to the gardens below. Never again
will there be resources to create such formalised splendour,
and the layout of the park beyond the house completes, for
me, the most beautiful spectacle I know. The park tails off
into primeval oak forest, the habitat of plants and wild
flowers which cannot survive except in such undisturbed
seclusion, so that naturalists fall as deeply in love with this

unspoiled countryside, left undisturbed for posterity, as others do with the artifices of more ornate designs.

The present Duke of Devonshire has such faith in the future that he plants about fifty young oaks a year to replace the venerable veterans which collapse and are returned to the soil. He plans for the heirs of his heirs to enjoy the same wild and natural forest for centuries yet to come.

Ten days or so spent filming with the owners and their staff offer many opportunities of assessing how they get on – and we never heard an unkind word on either side. Not only had ownership passed down generation after generation of the same family, but the same hereditary theme united as many generations of the estate workers. The Duke and Duchess of Devonshire knew them personally, by name, as well as they knew the Duke and Duchess. Although there was mutual respect, they all talked together with the uninhibited ease of old family friends. There were three villages on the estate to house estate workers. No houses were available for new staff for the simple reason that they were occupied by retired workers or their wives. Old employees are not chucked out at Chatsworth! What a lesson for the so-called Welfare State! What an advertisement for the best in country life!

There will, of course, be left-wing city slickers who will moan that such relationships are too feudal and paternalistic, that land should belong to the state, not to rich landowners, and similar trendy cant. The fact that whole families live and work on the estate for the whole of their long lives – and would hate to live anywhere else – puts such prejudice in perspective, because the pride in keeping the great estate alive was mutually possessive. 'We' plant trees for 'our' heirs – whether 'we' buy the trees or 'we' dig the holes to plant them. The interdependence was not limited to owners and workfolk because even the public were included.

The house and grounds are open to the public, who pay for the privilege. Even so, the cost of repairing and maintaining such an immense house and estate is astronomical, so that the income may go some way towards mitigating the loss but has no hope of turning into profit. But centuries ago, when the nobility were wallowing in riches and had no need of such chicken-feed to balance the books, it was still common practice to show perfect strangers the family treasures whenever they cared to ask. Families who have owned huge tracts of land for centuries are brought up, from the cradle onwards, to regard themselves as stewards for the future.

In commerce, it is common for the first generation of entrepreneurs to get a foot in the financial door, the next generation to make a fortune and the next to squander it. Hereditary landowners, on the other hand, grow obsessive about their land and are often prepared to make great sacrifice to leave it better than they found it. When they own vast tracts, such as Chatsworth, Woburn or Longleat, villages do not get taken over by jerry-builders, woodland is cared for, both as amenity and for hunting and shooting, and although hedges may have to be grubbed-out for modern agricultural machines, they are done so with sympathy, opening up the view, maybe, but with discretion so as not to spoil it.

The dangerous alternative is for large areas of land to be snapped-up by pension funds, universities or faceless Big Business, who are neither interested in those who have lived there for generations nor the land they tend. All that such owners hold dear is the profit and loss account appearing on their balance sheet. They put in agents, who operate with about as much security as football managers, highly paid when things go well, but out on their ears at the first hiccup. Such propertied prostitutes are motivated by nothing but profit and nothing is sacred to them if its sacrifice will make another bauble.

Each succeeding generation of the really great land-owners, on the other hand, usually specialised in a different way of leaving his land better than he found it. Some estates were laid out by Capability Brown, or his disciples, to give avenues and vistas, lakes and rolling parkland to fill succeeding generations' eyes with pleasure. Joseph Paxton, at Chatsworth, and his friend and employer the sixth Duke, left waterworks to posterity that cut the Russians down to size, to say nothing of well-treed park-land – where amenity trees are still being planted but primeval forest oaks are left in peace. Coke, of Holkham in Norfolk, was a pioneer of agriculture who left his indel-ible mark on the best of British farming, while the Earl of Aberconway did as much for the arts of horticulture on his estate at Bodnant, in North Wales.

Cynics, with green eyes, scoff at those who own what they can only envy. But the fact remains that it was the great landowning families who fashioned the cultivated countryside which is now threatened most by the agricul-tural revolution. The combination of vast wealth, good taste and almost unlimited leisure allowed the landowners, in centuries past, to commission Capability Brown, and whole schools of other skilled landscape architects, to lay out their estates to create visual perfection. The courses of rivers were altered and diverted to spill into eye-catching serpentine lakes and specimen trees were planted at conspicuous vantage points. If a village spoiled the view, it was razed to the ground and re-sited, not where it would be invisible, but where it would make a positive contribu-tion to the landscape.

Lakes and fountains, follies and grottoes, combined with great houses to delight future generations as well as present owners. Everyone living on the estate, man as much as master, was fired by the enthusiasm of personal involve-ment, the satisfying sense of shared achievement, untasted by the world of spurious sociologists, forever preaching about the virtues of some welfare state. It will be a sad day if Big Brother, some faceless Ministry of Agriculture boffin, ousts or replaces those whose families created and defended the pattern of the English countryside, because such bureaucrats are motivated only by politics and prejudice.

They are keen to point out that it is sentimentally uneconomic to retain stately trees in ancient parkland. Only improvident fools, layabout landowners and woolly conservationists kick against such pricks of progress. Trees get in the way of combine harvesters, sugar-beet lifters and mole ploughs. So they give subsidies to grub them out and, even if they are forced to lash out more subsidies to replace them later, it will take generations to restore the beauty that they desecrated. Left to such custodians, the countryside will have no sleek horses, idling away sultry summer days in the shade of ancient oaks or chestnuts. The rich turf of old meadowland will be stripped to make suburban lawns, where it will be sprayed with chemicals because wild flowers, in venerable meadows, are weeds in artificial lawns. When the parkland has been reclaimed for agriculture, it will be planted with corn, which will be sold at a loss to Communist countries, who are parasitic on our mountains of surplus corn and lakes of milk.

Estates which escape Bureaucratic Big Brothers may have Big Business Big Brothers wished on them instead. The Rolls-Royce-and-Runny-Nose brigade, with more money than manners, live under the delusion that a place in the country will be a status symbol that will secure their recognition by respectable country society. So when estates are broken up, after centuries under the stewardship of families who have dedicated themselves to the land of their forefathers, there is great resentment if they are acquired by some tycoon who thinks the brass he's made will buy respect. 'Twopence ha'penny, looking down on tuppence', does not cut much ice in rural circles and the local ambition is to cut the stranger down to size. The result, all too often, is that the newcomer gets fed up with rebuffs, so fills his house with visitors from the commercial world whose customs he shares. He may live in the country – but he doesn't belong. This gin-and-jag set, with their swimming pools and squash courts and extrovert symbols of success, change not only the face of the countryside but its character as well, because they are never respected as their predecessors were.

9

Down on the Farm

Farmer-bashing has long been a favourite blood sport among city folk and it has now degenerated into a growth industry with trendy media men. To some extent the agricultural moguls have only themselves to blame. Despite a flashy lifestyle, they have pleaded poverty so loud and long that there is now a temptation for the public to write-off quite genuine catastrophes as just another cry of wolf. Not many of us can remember the last time we saw a farmer on a bike.

Their trade inevitably fluctuates because seasons which are perfect for one crop may spell doom for another, and unseasonable weather can cause unpredictable failure or success. Every shortage inflates the price, tipping the jackpot into the silk-lined pockets of the lucky – or shrewd – gambler who happened to pull the right lever on the farming fruit machine. This didn't matter in the old days because the price was self-regulating. When it rose there was an unseemly scramble to join the bonanza, which resulted in prices slumping again. You can't beat the law of supply and demand except by artificial means. So the able farmers prospered while the weak went to the wall.

Two World Wars drove home the lesson that, since we live on an island, home-grown produce can make all the difference between survival and destruction. It was government policy to grow Food for Britain at almost any price so that for several generations farmers have been encouraged to grow the maximum amount of food and any loss they suffered has been made good – or better! – by grants or subsidies in one form or another. The mythical farmers' bike has been replaced by shiny BMWs and Range Rovers, while their wives swan around in swimming pools. A succession of second-rate Ministers of Agriculture, forming a coven with their counterparts in Common Market countries, have exacerbated the position by mushrooming subsidies until farmers have produced such mountains of surplus grain and lakes of surplus milk and bogs of surplus butter that the Common Market has almost gone bankrupt getting rid of the excess by selling at a loss to our Communist enemies. Many of these ministers had farms of their own so that, apart from being too wet to stand up and be counted on the side of common sense, they have had the incentive of lining their own pockets by voting for subsidies at taxpayers' expense. The last minister with the guts to shout unpopular odds was Stanley Evans, who proclaimed, about a quarter of a century ago, that farmers were 'feather-bedded'. It didn't do much good because he disappeared in a spray of expletives and his political epitaph says simply 'Exit Feather-bed Evans'. But he will be remembered longer than his insipid successors.

Farmers are still feather-bedded because rural votes in Euro-elections are as important, to equally feather-bedded politicians of the Common Market, as any votes in time of war. But the writing is on the wall because, quite simply, the funds for subsidising agriculture are running out. The truth is dawning that the law of supply and demand can only be thwarted temporarily and at an unacceptable price. The figures involved are very stupid. By the early eighties, we were forking out three and a half *billion* pounds to guarantee farm prices, which would otherwise have slumped because we had let supply exceed demand. On top of that we were spending about as much again to dump it on foreigners, including Communists, who therefore have that much more to spend on weapons to use against us when it suits them. In other words, we were squandering fortunes to produce an embarrassing surplus and doubling the loss to get rid of it. No wonder the public got cross! The green-eyed god of jealousy, which provoked hostility against farmers, should really have been directed against incompetent politicians, who devised the crazy incentives, which no one but fools would turn down. Sooner or later, the backers of this Gilbertian farce had to face the auditors and the first sacred cow to be condemned to be knackered was the subsidy on milk. A quota was proposed, with heavy penalties for exceeding it, so that milk producers who had previously had it so good stubbed their toes upon reality. They roared like bulls, switched political parties and changed their programme from producing milk to growing beef or corn, which were not immediately threatened. One does not need a very high IQ to predict that cows which had recently clambered out of the milk lake would soon be perched perilously on the pinnacle of a beef mountain. Any farm product which needs heavy subsidies to survive must also be on the list of fiscal changes imposed to balance the nation's books.

Produce in quantities to create surplus lakes and mountains also happens to be very bad for the appearance of the countryside. Farmer-bashers and the lunatic-fringe conservationists prate endlessly about farmers who create deserts of plough, where once there were pretty little fields, and grub out hedgerows that gave form and interesting pattern to the countryside. They complain that farmers drain wetlands, fill in pools where once frogs bred, and chop down woodlands planted by aesthetic ancestors. All of which is, sadly, true.

Before seeking a remedy, it is sensible to examine what really happened in the past. At present, four people out of every five live in towns and, in an era of insecurity, they are naturally bitter about the squalor and vandalism and violence of urban life. Their grandsires or great-grandsires may have migrated from the country during the Industrial Revolution, a century or so ago, so that they will have grown up with an idealised illusion of life on the farm as lived by their ancestors. They may have been in the Women's Land Army or their parents may have been evacuated to the country during the last war, or they may simply be sensible enough to prefer country holidays at home to being fleeced, in hordes, by foreigners. Whatever the cause, there is a great temptation to fall for the romanticised folk image of life on the land 'in the good old days'. They resent anything that threatens change. The fact is that the countryside has never been static, nor will it ever be. It always has changed. It is changing now and it will always continue to change. Time will not stand still.

The continuous growth of population down the ages makes this obvious. When William the Conqueror commissioned his Domesday Survey, there were one and a half million inhabitants, less than in one medium-sized commuter town today. The land was almost continuously clothed in woodland and it was said that a red squirrel could have swung, from branch to branch, from Land's End to John o' Groats without ever touching the ground. I expect that was an exaggeration, but the fact was that large towns, by their standards, were less than commuter dormitories by ours, and farmers needed only small clearings in the native woods to grow their crops and feed their stock. By 1377 there were two million inhabitants, over four million a century later and six and a half million when the first English census was taken in 1700. By 1850 the population of the British Isles had mushroomed to more than twenty-seven million. It was over forty million by the First World War and now it is over fifty-six.

So it doesn't need much imagination to appreciate that multiplying the population twenty-five or so times must entail altering the face of the countryside to accommodate and feed them. The greatest problem to be solved world-wide is over-population, which breeds insecurity and aggression, but whether we solve it by disease or famine, war or birth control, there is no way that we can cocoon the countryside from change.

Such changes have probably been more dramatic during my lifetime, which spans the last seventy years, than during any comparable span in the past. I grew up in the horse-and-cart era and amongst my own most nostalgic memories are sunny summer days spent haymaking on the local farm. In common with such idealised memories, the hard work of hand raking and loading hay wagons and the seeds that penetrated eyes and ears and nose and every crevice are conveniently forgotten. The times I remember are the pleasant ones when, as a kid of ten or twelve, the wagoner heaved me onto the back of the lead horse so that I enjoyed the delusion of being in charge of the team. I remember swigs of sweet cold tea from bottles – and forget that, if Thermos flasks had been invented, which I doubt, farm labourers were unable to afford them.

That was in the 1920s and, soon after, horses were replaced by tractors, even on small farms. Everything now happened at a greater pace, but corn was still ricked and threshed by a contractor, who moved on from farm to farm, trundling his great rattling threshing box behind a wheezing steam engine, which pulled it from site to site and then provided power to drive the machine by huge leather belts that could decapitate anyone standing near them when they broke. It was the noisiest, dustiest, dirtiest form of hard labour ever devised by man and modern shop stewards would go into terminal decline if they even saw a photograph. But looking back, through my rose-tinted specs, I never see the hard labour or danger or squalor. All that I remember clearly is the fun we had when we reached the bottom of the rick. The connubial produce of a winter colony of well-fed rats might amount to several hundred strong, which retreated as the rick was threshed, till they formed a rearguard in the wood-and-straw foundations.

My ratting dogs were better known than I was and there was always a welcome for them on any farm where threshing was in operation. The last few rows of sheaves were pandemonium, as we caught the rats when they attempted to escape. Hairy Kelly, the old professional ratcatcher, would have made modern so-called Pest Officers look the incompetents they are. No scientific gadgets for him. He caught them in his hands as they bolted and stuffed them, alive, into his shirt till a lull allowed him to transfer them to sacks, to sell in pubs to settle wagers about whose dog was best. My dogs were pretty good and two of them caught over a hundred rats at a sitting (or chasing!) and a thousand in twelve months. Considering I was at boarding school for nine of those, they didn't have much time to hang about.

Perhaps the greatest change of all to the countryside occurred when threshing machines and corn ricks, with beautiful thatched roofs, were replaced by combine harvesters and grainstores. Even the prototype combine harvesters made old-fashioned methods appear to have come out of the ark. In my day, three horses and a binder could cut and bind into sheaves an acre of corn a day. My neighbour's latest combine can do the job in less than fifteen minutes. With the old binder, the sheaves of corn were spat out behind the machine, which was followed by a gang of labourers who picked them up and stooked them in four or five pairs of sheaves, so that the wind could dry out the surplus moisture to avoid mould – at best – or spontaneous combustion when the corn had been taken and stacked in ricks near the farm. Months after that the threshing machine arrived to thresh what the rats hadn't eaten and build straw ricks for bedding to replace the corn ricks for grain.

Down on the Farm

Labour was cheap then, so the whole task was economic, but wages have gone up from 28/- a week, or £1.40 in our modern washers, to over £100 a week. So, although modern combines cost £30,000 to £40,000 apiece, the speed they work at makes the threshed grain far cheaper than it would be by the old labour-intensive methods. There are, however, snags. Modern combines and corn drills and chemical sprayers are so large and move so fast that it would be quite impractical to manoeuvre them in the little five- or seven-acre fields. Farmers liked them then because they were convenient to work and crops were planted in rotation so that weeds died out before the same crop was sown again. Chemical weedkillers and pesticides are now so effective that the same crop can be grown year after year on the same land.

In the old days, when arable fields were 'rested', they were put down to grass and relatively small fields allowed stock to be grazed and moved on, while the land lay fallow to break the breeding cycle of parasitic worms and flukes, before cattle or sheep were returned again. Nowadays my neighbour doses his sheep with worm medicine every three weeks – and runs them on the same ground far longer. So, small fields are no longer needed for stock and are a positive disadvantage for large machinery.

The new industrial revolution, with computers and automation and silicon chips, is far more responsible for modern urban unemployment than bad management or bloody-minded labour or economic slumps. By the same token, mechanisation of farm equipment (even individual dairy cow rations are now calculated by computer!) and chemical farming to control pests and weeds, are the main reasons for the large fields and diminished hedgerows that conservationists deplore. The result may be as bad, but the fact is that old-fashioned equipment and labour-intensive field layouts would send the farmer up the spout before next harvest.

What Price the Countryside?

Perhaps the most provocative by-product of modern agriculture is the practice of stubble burning. Our politicians and their French collaborators have encouraged farmers to grow as much corn as they possibly can and promised to keep up the price at a highly profitable level, however big the surplus mountain of grain their potty scheme produces. It has been artificially more profitable to sell grain at a loss to Communist foes than to feed it to our own cattle and pigs – so there have been proportionately fewer cattle and scores of pig farmers have gone broke. There has been no market for the tens of thousands of tons of straw in arable areas because it has not occurred, even to our Ministry of Agriculture, that farmers might enjoy a subsidy for creating a straw-bonfire to warm the milk lake and cast light on the grain mountain. As a result, straw is not only worthless but a positive obstruction to the harvest process. Farmers naturally have taken the easiest way out and set fire to it. Every harvest, for several years , the skies have been obliterated by a pall of acrid smoke from stubble fires. Storms of sooty smut have blacked out whole villages, ruining brickwork, washing, houses and health. The climax came when impenetrable smoke, thicker than pea-souper fog, drifted across a Yorkshire motorway, causing a horrific pile-up with tragic loss of life.

Public anger exploded, but the Ministry was too wet to make the practice illegal with draconian penalties, but relied, instead, on farmer co-operation with a voluntary code of practice. The inevitable happened. The few exceptional louts in the farming fraternity did not give a damn for public opinion and the good name of the vast majority suffered because of them. Public pressure continued and, as always happens when votes are threatened, the politicians eventually caved-in and paid at least lip-service to the protest by slapping a legal ban on the practice, though how effective it will be is still to be proved. The one certainty is that the reputations of the majority of responsible and innocent farmers suffer for the few guilty.

The real damage that stubble burning does is to wildlife. Every acre of stubble holds millions of insects and spiders and other minute wildlife, some of which are pests and others beneficial. Fire does not discriminate. The scorched earth policy destroys the pests and their predators, birds and hares and other animals are all fried alive and, when the holocaust is spent, there is a sterile desert, where even the worms which aerate and drain the topsoil are baked as inert as a gypsy's hedgehog baked in clay.

The Ministry of Agriculture cares nothing for such trifles. All that concerns them is the number of votes their masters could lose if the practice continues. So the straw-burning code stipulates the width to be ploughed as a firebreak between stubble and hedges and lays down that burning should cease on Sundays and Bank Holidays, when the maximum number of voters would be playing in the countryside and might vote the wrong way if their T-shirts were soiled. They do not give a damn for the fact that wildlife fries as lethally on weekdays as on the Sabbath.

Farm mechanisation, which cannot be practised without relatively large fields, has reduced the requirement for labour, precisely as automation has in factories. Men employed on the land have dwindled in the last few years and the output of crops, including milk, has spiralled. Since prices have been pegged fictitiously high, due to the gnomes in the Common Market, profits have soared and land prices with them. Only the very rich can now buy farms and even those who already own the land need astronomic capital, with tractors costing up to £50,000 apiece and implements in comparable brackets. Even men who have farmed in a very big way for years find difficulty getting their hands on cash for capital equipment so that more and more land is owned by institutions and pension funds, at the expense of the men whose hearts have always been in the land.

Private ownership becomes less and less possible because

when an owner dies, his successor is usually forced to sell part of his estate to pay death duties. Even when owners sell land in their lifetime, Capital Transfer Tax is so penal that land often has to be sold to find the money for essential repairs or development of the rest. As private estates shrink, they are snapped up by finance companies, corporations, insurance funds and City institutions, all of which have one thing in common. They do not die. Even 'conservative' governments stack the fiscal cards against private ownership in small, personal units. The sad modern trend is for more and more land to get into the hands of fewer and fewer *impersonal* 'owners', whose only yardstick of success is return on their capital, whatever happens to the landscape or wildlife.

Many of the new agricultural landowners sit in city offices and wouldn't know a bull from a stallion if they mounted an expedition into the rural hinterland. They farm the land with college-trained farm managers, whose success or failure is judged solely by the balance sheet. The job security of such men is comparable to industrial executives and they have to maximise profits, at any cost to the countryside – or else! Literally thousands of small farmers have been squeezed out of business by this stupidity of rigging prices and the tycoonery of far-off businessmen, who admire the countryside only in terms of no more than profit and loss. A return to the basic law of supply and demand, which must ultimately rule the market place, would deflate the profits which attract Big Business and at least slow down the rise in land price.

Nothing, of course, will ever turn back the clock of mechanisation, for that is a common factor in industry as well as farming. But there have been recent signs of an upsurge in agricultural contracting, where specialists own machinery, not land, and contract to harvest, harrow, plough and sow for many farmers in their locality. A large combine harvester, capable of cutting and threshing

78

perhaps an acre in ten minutes, may deal with the whole crop on even the largest farms in a couple of weeks or so. The machine may cost £40,000 or £50,000 and even the biggest big boys don't like writing off such capital sums in a couple of weeks' productive work a year. If a contractor owns the same machine, he can get far more work out of it by going on to the next farm as soon as he's done the first. It will do nothing to reduce the necessary size of fields, of course, but at least it will allow smaller farmers to get in on the act because they will need so much less capital to start.

If there is anything that can be done to encourage the men who farm the land to live there, either as owners or tenants, the countryside could revert from huge, impersonal blocks of land to the cared-for countryside so many people yearn for. The old country saying that no muck is as good for the land as the sole of the farmer's boot is as true today as ever it was. Men who are bred and born and reared on the land they farm appreciate everything about it and, unlike the absentee businessman farmer, they will not prostitute themselves by degrading it for a few pounds extra profit. More farmers like that would mean less to provoke controversy.

It is pointless for the lunatic fringe of conservationists and environmentalists to mount strident farmer-bashing campaigns because all they do is to alienate struggling farmers more than the real rich culprits who sit, aloof and unscathed, in city offices, leaving their harassed farm managers to carry the can. The chaps to train the sights on are the politicians, who lash out largesse, in the shape of subsidies and capital grants, which can only make the rich richer.

My job, as a country writer and broadcaster, throws me into contact with landowners and farmers all over the country and the plain fact is that men who have grown up on the land, working and playing on the patch their ancestors worked and played on before them, have a far deeper love for it than the superficial nostalgia of strangers who only visit it on sunny Sundays. It is useless to try to roll back the physical reality of mechanical progress because that is with us, whether we like it or not, and it will continue to develop as men make new inventions. So it is short-sighted to alienate the majority of real farmers because a few exceptions exploit the land, for such campaigns of farmer-bashing simply drive the wedge of conflict deeper between the vast majority of sensible townsfolk and countryfolk. The real culprits are the politicians and bureaucrats whose efforts to create a phoney cloud-cuckoo-land have no more chance of success than Canute had of turning the tide.

10

Rent-A-Farm

Like it or not, the rural mechanical revolution is here to stay, if only because it has taken so much of the unpleasant irk out of work. Tractor drivers who now sit in comfortable, heated cabs have mechanical noise blanketed out by earmuffs, connected to radios which relay the blarney of Terry Wogan, the dulcet tones of sexy pop stars or incoherent racing commentaries, at the touch of a button on the tractor dashboard. They would look down their noses at an invitation to plod along the furrows behind a horse-drawn plough again – even to preserve the aesthetic pattern of little hedgerowed fields. The idea of hand-milking twenty cows would be equally unwelcome to the chap who only has to act as sergeant-major in a milking parlour where machines do the same job five or ten times quicker, with labour limited to fitting rubber caps to bulging teats.

A more profitable target to aim for is the stupid legislation which has made it virtually impossible for any but the very rich to get a farm. In the heyday of farming there were a great many tenant farmers because landowners were only too happy to divide up their estates into workable farms and lease them to tenants. It was perfectly possible for the owner to repossess the farm at the end of the lease but, in practice, this rarely happened, except where an idle or inefficient tenant let the land deteriorate through bad husbandry. In the majority of cases, a close and friendly relationship matured between owner and tenant so that it was common to come across farming families who had been tenants of the same landowners for many generations. Until the outbreak of the First World War, more than ninety per cent of all farms were tenanted because the great landowners owned so much of the country. Some of this land was owned by the Church, some by the Cown and a great deal by the descendants of noblemen who had been rewarded for service to king or country in centuries past, or whose women had been royally bedworthy. The home farm on most estates was farmed for the owner by his bailiff, but the rest of the land was let to tenants whose family tenure of the land was often almost as long as the owner's.

The Great War and the depression that followed it combined with death duties to disintegrate the rich estates so that a great deal of the land was sold. Unfortunately many farmers went bankrupt so that comparatively few tenants were able to buy the farms which they and their forebears had cherished for generations. Some of the prices seem utterly absurd, by today's standards, because land that went for between £20 and £30 an acre in my young days would fetch in excess of £2,000 now. Even so, inflation made it relatively more profitable to sell land than to let it and the Labour government's stroke of lunacy was to pass a law in 1976 to give tenants security of tenure for three generations. This blunder converted the shortage of tenanted land from scarcity to famine as, in inflationary times, nobody but a fool would let land if he could help it because the price of rented land trebled between 1972 and 1982. As tenancies fell in, through death or other causes, the vacant land was either sold or taken into the landlord's possession to be farmed by his agent. Present land prices give a return on invested capital far higher than rent obtainable so that various ways have been investigated to make letting more attractive – by limiting security to the lifetime of the tenant or by fixing rent as a proportion of what profits independent assessors calculate the farm should make. Taking into account the unpredictable effects of weather – which will vary in different parts of the country – and the impossibility of defining an average farmer, it is small wonder that progress is minimal.

The decline in farms available for letting had been caused through the uncontrollable pressures of the slump and by the stupid socialist policy on tenancies, so that 2,000 small farmers a year were forced to leave the land. They were replaced by farm managers, who are forced to dance like puppets on financial strings, pulled by the new institutional owners, which has been the cause of many of the ills attributed by the farmer-bashers to small farmers and moguls alike. Individual branches of the National Farmers Union campaigned for higher capital grants and tax relief for small farmers and a code of practice to prevent landlords amalgamating farms into too large blocks. A strong body of opinion claims there are no good reasons for having farms larger than about 500 acres. When the plea was published, the headquarters of the NFU disagreed, announcing that fewer and larger holdings were seen as 'dynamic' and 'competitive' so that they would do nothing to limit the size of holdings. Since subscription to the NFU is on an acreage basis and the institutional membership is high, this will not come as any great surprise because many cynics have long thought that farmers need no enemies with friends like the NFU! The president, Richard Butler, who farms 1,400 acres in Wiltshire, claims that less than twenty per cent of land will be tenanted by the end of the century, in which case it would seem unlikely that any but the big institutions could raise the wind to buy it. If, as president, he is prepared to concede this, it is surely time for a shake-up.

So, if the farmer-bashers wish to be constructive instead of simply counter-productive, they should take care not to generalise. There is no doubt that many common farming practices are extremely destructive to the countryside, but a great many of them are caused by pressures of an artificially rigged market and stupid laws that are denying the type of medium and small farmers, who have traditionally been true guardians of the countryside, the chance to farm there. At the moment, the nation is spending vast sums in grants and subsidies to encourage the maximum production of food at almost any price, and vast sums more to dump it abroad to keep prices high at home. Far better, surely, to produce what the land will give *economically* and to spend some of the savings on providing conditions where owners and tenants can thrive more easily than the moguls. The smaller farms would automatically create a more beautiful and varied countryside, because men who live on the land they own or rent are so much more likely to cherish it than impersonal pen-pushers in far-off cities.

The last straw is that landlords have a nasty habit of dying – which institutional owners do not! Death duties then force the sale of yet more land at a price which only Big Business can afford.

A countryside, farmed by impersonal land agents for faceless institutions, interested only in profit, would fulfil the gloomiest predictions of conservationists that the aesthetic and wildlife aspects in the countryside are in severe danger. But suppose conditions were organised to enable small farmers, either tenants or owners, to set-up in small units. How would they fare in the new agricultural revolution? The capital cost of farm equipment is now so high that tractors can cost from £12,000 to £20,000 or more apiece, combine harvesters double that and other implements in proportion. Many of the more specialist machines would eat the work on a small farm in a few days and stand idle in the shed for ninety per cent plus per year. Small men cannot finance such outlay for such a short return. But if much of the mechanical work was done by agricultural contractors moving from farm to farm the capital outlay could be spread over the same area of land as if one financial whizz kid owned it. It is perfectly true that such mammoth monsters would need large fields to cavort around, so that the conservationists would still whinge about hedgerows being grubbed out, but it is never possible to stop the clock and freeze mechanical development – even if ploughmen could be persuaded to plod their weary way behind horses when the big fields were cut down to size again. At least the men who lived on the land, and loved it once more, would see to it that their fields were kept a sensible size instead of striving for the maximum.

Perhaps the fact that small men have to live more from hand-to-mouth than their Big Brothers is not a disadvantage. The volume of output, on their smaller scale, would be less than the mass production of intensive husbandry. They wouldn't produce the surpluses if they had to meet normal conditions of supply and demand, so subsidies for producing quantity at any price would diminish and, with them, the absurd cost of flogging surplus lakes and mountains at a loss. Some of the economies achieved could be ploughed into encouragement to produce quality instead of sheer quantity. Free-range eggs, meat produced from animals whose growth is not forced by hormones and other additives in their diet, would not only make a pleasant change for us, it would provide pleasanter conditions for livestock. Small British farmers, after all, were the best stockmen in the world who, for generations, produced breeding stock of a quality no one else could touch. If they were helped off the treadmill of blind mass production and allowed to use their traditional skills, their stock and their customers would both benefit.

To a lesser degree, the same applies to the small arable farmers. The barley barons have turned vast acreages of fertile soil into featureless prairies of plough which is a monoculture of corn as boring as any softwood forest. They souse it in chemicals, often spewed from sprays in aeroplanes, from first growth in spring to harvest in autumn. Then they set fire to the straw, blanketing the countryside in smoke and smut as filthy as the scorched earth of foreign battlefields. Small wonder that they are bashed by city folk, pining for the glories of our rural ancestors. They are hated no less by the unfortunate locals condemned to live downwind of their filthy conflagrations. But why should they care? We live in a throwaway society, where everything from old motor cars to plastic milk bottles are disposable, and it would cost more to utilise the straw and stubble than it does to burn it. In any case, the decision makers who evolve the policy rarely choose to live in the deserts their greed creates.

Small farmers are different. For one thing they can't wait for their income to land on their breakfast table only once a year after harvest. So they are more likely to practise mixed farming and rear some stock that will breed or be ready to kill till the time their bank statements take on a rosy hue. The advantage of this is that stock and straw go hand in hand so that, instead of disappearing in choking smog, their straw has a chance of providing comfortable bedding for well-kept stock and returning to the land as well-rotted muck to restore fertility for later crops.

Opponents of such traditions point out that the output of food might fall but, if it did, taxpayers would have less subsidies and deficiency payments to meet. One argument claims that the beauty of our land and the welfare of our native animals and plants are of no account because any surplus we can create should go to feed the starving masses of the Third World. This may well be true, in the short term, but it seems to me that the growth of world population is the greatest threat of all. Control of population growth, by war, famine, disease or active management is ultimately unavoidable – and the least painful is surely some form of positive birth control.

So a shift in government policy, to encourage smaller

farms to produce high quality food, under conditions that allow stock to live a dignified and comfortable life, would also have the effect of restoring much of the attraction of the landscape around which so much conflict has raged.

11

The Factory Farm

The modern obsession for quantity, even at the expense of quality, has diluted the ancient art of stockmanship more than any branch of husbandry. For countless generations before the development of antibiotics and all the abracadabra of modern scientists it needed great skill to keep livestock healthy. Old fashioned farmers knew all about the dangers of overstocking and getting their land contaminated with diseased animal droppings that could infect the stock that followed. They avoided such land sickness by rotating crops so that the interval before the next stock grazed the land was longer than the life-cycle of disease the last stock left. Sheep were constantly driven to fresh grazing on the hill, to rest the land where they last fed. Cattle were paddocked in fresh pasture by rotation and, above all, numbers were kept down below the danger level of overstocking.

Scientists have altered all that. 'Zero-grazing' is their euphemism for the system where cattle can be constantly housed and the grass brought to them by tractor, to be fed in troughs. They need never go into a field at all! Beef cattle are fed hormones to stimulate abnormal growth – and the fact that their flesh has the flavour and consistency of a sweaty sponge is of no account because the clever salesman can persuade gullible customers that it is lean and tender – and the convenience cookbooks will always prescribe some continental sauce that will impart exotic flavour – though not necessarily of beef!

Sows are kept in stalls, so narrow that there is not even room for them to turn round, and they are never let out for months on end till their time is come to be served by the boar again, to have another litter. Living in such skeleton coffins allows them to be stocked as dense as women at the cash till of a supermarket. The piglets they breed are crowded into sweat houses, where the temperature is kept high enough by their own overheated bodies, to ensure that every scrap of food they eat goes into growth, not keeping warm. They have no light but artificial, which is sometimes only switched on, at intervals, to allow them to feed, to give them less scope to do each other injury from frustration. Even so, they bite off each other's tails from boredom, unless the stockman(?!) amputates them first and gives them hanging chains to chew to pass away the time until their execution.

Poultry are kept in battery cages so small that they cannot stretch their wings – which would be illegal for any bird purchased from a pet shop. The conditions of broiler chickens for the table are equally bad. The stench in such places is overpowering, clinging to clothes and hair of operators (*not* stockmen!) who work there till they themselves are socially unacceptable. The latest craze is 'bio-mittant' lighting, and the Ministry of Agriculture is enthusiastic about research at Reading University where boffins have discovered that growth and production can be maintained by very short, intermittent periods of light, so that electricity can be switched off, for forty-nine minutes in every hour, and, because hens are less active in the dark, they also require less food. Sixty-nine pence per bird can be saved in a year. The cost in dignity and comfort of the flock is callously ignored by the men from the Ministry!

The stock answer to complaints about such practices is that 'the housewife wants cheap food and old-fashioned methods cost more'. Perhaps that might be true in times of war, when the lowest quality food is welcome for survival but, as things are today, unrealistically high prices are paid to produce food, which is then sold to foreigners at a loss. So it might be better to grow a bit less and use the money saved by avoiding surplus to subsidise the selling price and to provide tolerable conditions for the stock. It is no coincidence that customer resistance is building up against the worst malpractices of the new factory farming, partly on grounds of cruelty and partly poor quality.

For many years calves were taken and reared, singly, in 'crates', which were precisely what the name implies. A wooden crate, like a sow stall, was provided for each calf which could stand up and lie down but had insufficient room to turn round. It was reared in this until large enough to kill for veal, often being fed on food from which iron was deliberately excluded, to make the calves anaemic, which produced the white flesh the housewives liked. Such practices, devised not by farmers but by scientists, 'educated' at public expense, eventually produced the inevitable back-lash of public abhorrence and sales resistance. Some of the biggest veal producers were forced to abandon crate-rearing and keep calves in strawed yards, which at least was some improvement.

By the same token, the battery egg men are coming under pressure to sell free-range eggs from birds which at least get some freedom, some daylight and room to spread their wings. The bureaucrats are being as obstructive as possible by refusing to define free-range – and how much space it needs to wander before a hen can be said to have its liberty on free range. This, of course, plays into the hands of the spiv poultry farmer, who keeps half a dozen moth-eaten hens scratching about the farm yard, to create the illusion that they lay all the eggs he sells at the extra price the public are prepared to pay for natural eggs. The fact is that he will have a battery house round the corner, where 99.9 per cent of the pallid eggs he sells are laid.

The Factory Farm

Pressure groups of bearded weirdies, waving banners and whingeing strident slogans do nothing to improve the situation. They are simply written-off as militant nutters and they bring discredit to genuine conservationists. The effective way to improve conditions is to press members of parliament to divert financial aid from quantity to quality on the farm and to do anything in their power to widen the fashion for natural food.

Among the greatest problems with intensive husbandry is disposal of dung. Long gone are the days when stock cattle were bedded at the beginning of winter on sweet straw, to which layer after layer of fresh straw was added as fast as the last layer was fouled. By the end of winter the cattle were standing on a gigantic, sweet-smelling manure heap of healthy dung and rotted straw. It was ideal for spreading on the land before spring ploughing and it produced the healthy aroma that smells sweeter to true stockmen than any seductive perfume in women's fashion shops. Cattle are now kept on slatted floors, through which dung and urine fall into slurry tanks, to be pumped out, eventually, and spewd in a foul cascade, as pungent as rotten socks, across the countryside. Countrymen, brought up on the harsh realities of rural life, mutter about new-fangled inventions that make sweet pasture smell foul as a sewer, but incomers, hived up in village dormitories, from which they commute to cities, are not so tolerant. They take out law suits and arrange for protest meetings with the parish council – and I have to admit my sympathies are with them.

The thrust for mass-produced food, to cash in on subsidies, will be irresistible till politicians get round to subsidising quality instead of quantity that escalates to surplus.

Antibiotics are included in the diet of stock, to guard against the diseases modern farming fosters – and the result is that factory food is laced with the residue, which can produce immunity to antibiotics in people 'nourished' on factory foods, causing disaster if the drug is then urgently needed in medical treatment. Some hormones used to promote growth in animals are quite as bad. A recent fatal outbreak of *salmonella* poisoning has been ascribed directly to the flesh of beef cattle, which produces strains of *salmonella* resistant to antibiotics because it has been fed on growth-accelerating hormones and antibiotics. I once did a favour for the principal of an agricultural college and he invited me to choose a favour from him in return. I thought of Henry VIII and the luscious capons that were used at feasts by his court. A capon is a castrated cockerel, which goes on growing when its urge for sex is gone, producing incomparable, succulent, tender flesh, despite the great weight of the bird. My friend said my request was easy to satisfy. The modern method was to implant a sex hormone pill under the flesh of the neck, which was absorbed into the body, destroying the symptoms and desire for sex. The treated birds did not crow, fight nor chase hens but they grew as large and flabby as human eunuchs. Almost as a throwaway, my friend remarked that the practice of inserting the pills below the flesh of the neck was relatively new. Until recently they had been implanted in the breast. I naturally enquired what had triggered the change in procedure and my friend explained that 'the pills take several weeks to be thoroughly absorbed and it is common for the carver to reserve for himself the best cut off the breast. If this happened to include the unabsorbed remnants of the pill, it was apt to sap his pleasure for the following weekend or so!' I decided that I would manage without capons because the surgical method of castration used in Henry VIII's day is an intricate operation which is now illegal. So, if capons are offered on the menu of expensive restaurants, it is worth making a few enquiries – or choosing another dish.

Unfortunately, drugs and hormones of various sorts are commonly used in intensive husbandry, and, while not all are dangerous, they tend to produce the tasteless pap that needs camouflage with sauce or spice to have any flavour at all. And the other marvel of modern science that gets farmers a bashing is the practice of injecting meat with water to make it succulent. It may be no coincidence that the added worthless weight, contributed by the water, fetches the same price per pound as genuine meat. The blame for such malpractices should be laid at the door of scientists, who invented them, or retailers who exploit them, rather than at the door of farmers.

The end product is the same because a disenchanted public have caught the craze for organically-grown natural produce. Unfortunately some of the health food fanatics are anything but objective. I was involved with a radio programme about organic farming, some time ago, with one of the leaders of the movement. He gave us a splendid lunch of home-grown lamb, fed on pasture that had been defiled neither by pesticides, herbicides nor artificial manure. I noticed, on the pasture outside his dining room window, more molehills than pimples on a skinhead. More to make conversation than anything, I said that I presumed his 'natural' farming methods had attracted them. 'Yes,' he said. 'The compost and farmyard manure attract the worms and the worms attract the moles.' 'What do you do about them? Trap them?' I asked. 'No,' he said, 'I get the Pest Destruction Officer from the Ministry.' Now ministry ratcatchers were once skilled men but it is impossible to persuade respectable people to do the sort of work required now. Ministry pest officers cannot trap moles, in the old skilled way, but descend to dipping worms in strychnine and leaving them in the mole runs. So my health-food friend jibbed at having artificial manure on his land but looked the other way when the most diabolical and persistent poison was put shallow enough to be ploughed to the surface next season!

Amused by my discovery I could not resist the temptation to pull his leg about his wholemeal flour. He was very proud of his flour mill, which was as traditional as time, except that it was powered, not by wind or water, but by electricity! It ground wheat between two mill stones, rocketing its price two or three times. I had understood that 'dressing' the stones, to make them grind correctly, is the most skilled job because grooves, precisely the right shape and size, have to be chiselled in the grinding surface of the stones, so I asked how he'd found an experienced miller. 'I didn't,' he said. 'I sent my chap away on a course.' The stones weighed three hundredweight new and when they had worn till they only weighed a hundredweight, they were no longer heavy enough to grind the grain, so had to be replaced. They usually lasted nine months, by which time two hundredweight of grit had worn away, but it hadn't occurred to him that he sold two hundredweight of grit, mixed with the flour, every nine months. No wonder wholemeal bread makes the bowels work so well!

Nevertheless, it is heartening to know that the pressure of public opinion can be extremely powerful when applied sensibly. CAMRA, the chaps who object to the clam-belly fizz that has for long been foisted on the public as good beer, have forced the brewers to reintroduce 'proper' draught beer as well as – or even instead of – their gassy keg rubbish. Sales resistance to battery eggs and crated sows is just beginning to be effective but there is still one malpractice which seems to infuriate countrymen more than townsfolk. When effective and difficult campaigns have been waged to ensure that stock we have cherished will die a humane death, it is indefensible to allow foreigners to flout our law by slitting their throats and bleeding them slowly to death without pre-stunning. Lorries are crammed with open crates, stuffed solid with live hens which will be sold, still alive, to customers who will take them home, to be subjected to archaic ritual slaughter. The fate of cattle and sheep is less obvious, because they are taken in clòsed cattle trucks to abattoirs where our own cattle are slaughtered, under supervision, by humane methods that have been devised to meet twentieth-century standards. But, in these same abattoirs, other cattle are held in 'crushes', turned struggling upside down, while their throats are slit to bleed them to death. Temperatures in the East and Middle East are so high, and the air is so humid that meat would soon have gone bad in the days before refrigeration. So religious laws were formulated to ensure the maximum blood was abstracted at slaughter, and only certain joints consumed, the rest being discarded as 'unclean'. Although refrigeration and hygiene have made such ceremony obsolete, many of our cattle are still subjected to this misery and, to add insult to injury, the joints rejected as too unclean for foreigners are sold to our butchers so that we actually subsidise the unacceptable cruelty and eat their rejects. If foreigners are not prepared to abide by our civilised laws, they should go back to their native lands.

There has been much recent publicity about three relatively new forms of stockmanship, fur farming, fish farming and deer farming. Fur farming, in the form of raising mink, could in theory be carried out as easily on any patch of city wasteland as in the deepest countryside. All that is involved are rows of small wire-netting cages, about three feet long and eighteen inches square, in each of which a mink is kept until its coat is mature and in prime condition. It is then slaughtered and skinned and the skin is cured and sold for women's furs. This is particularly risky when the farm or 'ranch' is in the country because mink ranches have recently become prime targets for militant nutters, who bring genuine conservationists into disrepute. These stupid vandals break into ranches, cut the netting and set the animals free to roam the countryside. Quite recently they 'liberated' fifteen hundred mink within ten miles of where I live. Mink are like large, extremely active ferrets, and have the added advantage that they can both climb and swim with great agility. They are exceptionally fierce predators which will tackle anything from the size of a duck or weakly lamb down to mice, voles and fish. Because of their love of living near water, kingfishers and sand martins which nest in holes in the bank are particularly vulnerable, as are game and domestic poultry.

A few escaped in Devon and the Pennine area some years ago and Ministry of Agriculture pest officers were commissioned to catch them. They wasted a great deal of time and money but were far too incompetent to succeed and eventually gave up. Colonies of mink bred in these areas have created havoc with native wildlife, as most introduced species do. This vandalism by protest groups could result in the establishment of a most efficient predator, which has no natural enemies here, and could destroy vast numbers of vulnerable wildlife.

It is difficult to probe the miniscule minds of such activists to discover their reason for liberating mink. They say it is because they believe it immoral to breed animals for fur. Presumably they are vegetarians, who creep round without wearing shoes, or are simply jealous of the rich, who can afford to give their wives or popsies fur coats, or of those prepared to work hard enough to earn the money to pay for them. In either case, women wore mink long before the fashion for fur farms and their mink were trapped by the leg in steel gins, where they might languish in agony for days, in the wilds of Canada or Russia, before being put out of their misery. So, if the present gangs of vandals succeed in making mink farms uneconomic, the mink coats of the future will come, once more, from foreign trap-tortured animals, instead of from those reared in the relative comfort of normal animal husbandry here. The damage such people do on the premises where mink are bred and reared is as nothing to the irreparable harm to wildlife the establishment of colonies of wild mink would do.

Deer farming is also a growing rural industry. Red deer are most commonly reared but some farmers also specialise in fallow dear, which are husbanded like cattle in fenced enclosures. Deer in the wild are exceptionally disease-free, mainly because they range over such wide areas, where keep is sparse, so that any parasites they evacuate with their droppings are likely to perish before they infect other deer. Experience in the relatively crowded conditions of deer parks has shown that they are susceptible to liver fluke, various internal parasites which affect sheep, and occasional cases of tuberculosis. There is no real bank of experience to indicate how they would fare under conditions of intensive management but it is reasonable to suppose that they would need similar preventative medication to other farm stock. But deer are still basically *wild* animals and extremely highly strung, so that nobody really knows what effect upon them the relatively constant handling of domestication may produce. The danger is that such intensive husbandry may induce diseases which may be passed to adjacent wild stock where treatment is impossible. This would have a most serious effect upon one of our largest and most attractive wild animals – with which it is rather stupid to meddle for the sake of an easy buck now and untold trouble later.

Fish farming is much more widespread, much more predictable and, in my view, as great a threat to our native fish. Brown and rainbow trout are exceptionally easy to rear, though rather specialised to breed. Some farmers breed the trout and sell on the young fish to restock 'put-and-take' fisheries (which will be dealt with on p. 143) and to fish farms who sell fish by the pound to shoppers. The fish are fed by throwing bucketfuls of specially prepared pellets into the water and they surge to the surface and eat them greedily, making the water boil with activity. The feed pellets look precisely like the laying-pellets sold for hens, but they contain a carefully devised mixture of protein and other ingredients formulated to promote very rapid growth. They produce horrid white, cotton-woolly flesh, so a substance is added to produce pink flesh which simulates the colour – but not the flavour – of wild fish. The conversion rate from purchased food to saleable flesh is very high, and since a strong current of running water from the local river or stream is directed through the fishery, very high stocking rates are achieved. Literally thousands of fish can be reared in comparatively small tanks.

Unfortunately much of what goes in as food comes out as excrement so that a fishery feeding several tons of food to fish a week, sends several tons (minus only the weight put on as growth) of excrement downstream. This is an extremely rich fertiliser so that it will change the density and type of weed downstream, destroying the natural balance of plants.

Fish are also as liable as other species to contract disease when overcrowded – which is another term for kept intensively. Spores of disease and infected excrement contaminate the fish downstream and one of the side effects of the latest craze of fish farming is that there is real danger that our native fish, especially salmon and trout, will contract disease swilled downstream with waste food and excrement from intensively-managed fish farms.

A fringe peril is that when fish fry are reared in unnatural numbers, the natural predators of fish, including herons, kingfishers and dippers, converge on the farm as tits would feed from a well-stocked bird table. They are then regarded as pests, as harmful as a keeper would reckon crows could be to pheasant eggs and chicks. Although the Wildlife and Countryside Act protects – in theory – kingfishers and herons, it is possible to obtain a licence from the Ministry of Agriculture to kill them if damage to property is proved. The Nature Conservancy Council has to be consulted but this is a mere formality.

Only this spring, a local fishery applied for a licence to kill cormorants which were eating trout at Blithfield Reservoir, near to where I live. Blithfield is not a fish farm but a fishery, operated by a club solely concerned with the sport of catching trout. Fry are purchased from farms and grown on in tanks, fed on pellets, until between 1¼ lbs and 3½ lbs in weight, when they are tipped into the reservoir for club members to catch. The fish, when set free in the reservoir without experience either of predators or catching their own food, swirl up to feed when anything, including an artificial fly concealing a hook, hits the surface. About 40,000 fish are put into the fishery each season and it would be an exceptionally awkward angler who failed to catch them. Cormorants are not awkward – so they make a living from the sport of their human benefactors. The sort of status-conscious angler who can afford £380 a year to catch a few tame trout did not make his pile of brass through being sentimental, so an application for a licence to kill the cormorants (protected by the Wildlife and Countryside Act) was made to the Ministry of Agriculture. The Ministry asked the Nature Conservancy Council for their comments on the application but did not even bother to wait for their reply. By the time the NCC had examined the case – and advised *against* any licence to kill the cormorants – it was too late. The Ministry had pronounced their arrogant sentence of death!

No department is as callous of the welfare of the countryside as the Ministry of Agriculture and it is quite wrong that they should be entrusted with the fate of any wildlife, rare or common. It would be far better for the Ministry of Agriculture to be restricted to giving advice and for the Nature Conservancy Council to have the last word on control because they are the only government body to be entrusted with the responsibility of *conserving* wildlife, as opposed to obliterating it. But as things are, the Ministry is only entitled to issue a licence to destroy protected species if 'serious damage is being done to food or property'. They should have no authority to give a licence to destroy birds which are only interfering with prospects of a very artificial sport and pose no threat whatever to the nation's food.

12

The Nation's Playground?

Farming has no monopoly of the countryside, nor did it in the past. In Norman times, the total population of the country was a million and a half, no more than a modern, medium-sized town. This tiny population was concentrated in small groups, which farmed common fields near to villages and most of them were employed by great land-owners, the nobles and the church. Vast areas of woodland, interspersed by marshy wetlands, stretched from Land's End to John o'Groats. Hunting was then the main preoccupation of the landowners, and bishops and soldiers competed ruthlessly for the sporting rights. If small things like villages or farms interfered with the quality of the chase, they were obliterated and laid waste. Commoners had various rights in the forest, such as the right of pannage, when they were allowed to turn pigs loose to fatten on the acorns. They could take wood and turf for fires, and timber for their houses. But woe betide them if they took a deer or other game, or even entered the wood during June, the 'fence month', when they might disturb the breeding docs.

By 1500 the population had trebled, and farming had grown far more important. It was, as yet, impossible to keep meat over winter, except by salting it, and so country markets had to be supplied with fresh meat 'on the hoof', which might have been driven miles by drovers using traditional tracks, or Drovers' Ways, to move cattle sometimes hundreds of miles to market, travelling slowly enough for them to feed on the way.

As the population increased still further and farms ate into the forest, small communities needed to go to local villages and towns to market, school and church. These communities were so self-contained that everybody knew his neighbours so that it was perfectly natural to make them welcome to take short cuts across one's farm or estate. They, in return, would have considered it a most unneighbourly act to cause any damage to fences or crops so that, when the land was clear, they would take the nearest route and, when crops were growing, they naturally kept to headlands and field boundaries. The custom became so habitual and the paths became so traditional that they were eventually regarded by all as legal rights of way, some being simple footpaths and others bridle ways, where riding or packhorses travelled. Most of the traffic on them was light, because the population was still so sparse, so that a right of way across one's land was no hardship, especially as its use would be mainly confined to familiar friends and neighbours.

These ancient rights of way still exist, enshrined as dots upon the Ordnance Survey maps, but they are now a source of endless friction. Conditions have changed immensely since they became an accepted pattern of the countryside. The population has trebled or quadrupled or more since men first trod them so that farmers and landowners are more likely to see strangers than neighbours crossing their land. Most important of all, field sizes and farming methods have been changed as much by the new industrial revolution as any engineering factory in the land. Where fields have been enlarged, to cope with new mechanised farming methods, paths cross the middle of fields instead of skirting them. This would not matter if the modern footpath users were prepared to move with the times to keep pace with modern farming methods, but all too often, they are not. Once a path has become a legal right of way, small groups of extremists and hard liners fight every proposal to divert it to cope with changing conditions. If a path crosses a field of corn, some of this lunatic fringe will vandalise it by treading a path through the growing crop rather than walking round until the harvest. One of their bogeys is a bull in the field. They try to make it illegal to put a bull in any field crossed by a public path, presumably because they simply do not know – or won't admit – the biological facts of bovine life.

The gestation of cows is nine months and they come on heat roughly once every three weeks, without giving any obvious physical sign that can be spotted by the cowman from a distance. To make things all the more difficult, the period for which the cow will stand to be mated by the bull is only a matter of hours. No serious problem is posed with dairy cattle, that are brought to the homestead to be milked twice a day, because there the cowman can pick out cows on heat and put them to the bull before returning them to the pasture, or phone for the artificial inseminator to come at once with stored semen from a chosen bull. Beef cattle, on the other hand, are kept out at pasture for weeks or months at a time, often far from the farmstead. The only way of ensuring that they breed more beef cattle is to pasture a bull with them, so that he can notice cows immediately they come on heat and serve them. If they fail to conceive, he is on the spot to remedy the defect next time.

Fortunately it is the milk breeds that have dangerous bulls, while Herefords, Aberdeen Angus, Red Devons and other beef breeds are normally as docile as their cows, however fearsome they may look. In the old days, the standard practice of countrymen who found a bull in a field was to avoid it by going round or, if it was a beef breed, crossing quietly because it was well understood that a bull running with the cows was vital for beef production.

Strangers, especially urban strangers, who do not appreciate that bulls are necessities not luxuries in such circumstances, go for the nearest policeman or even take legal action. The fact that they are equating their pleasure with the farmer's livelihood means nothing to them. They are happy enough to bash him for making larger fields to suit modern machines, but they are sensitive when he replies that they are equally selfish in resisting logical diversions to meet changing conditions.

The modern disease of teenage vandalism has made it desirable to rationalise footpaths that go through allotments and gardens – there was a recent report of rape at a girls' boarding school where it had proved impossible to have public footpaths through the grounds diverted to cope with the decline in public order.

Farmers do not all take the provocation lying down. One man keeps a colony of bees near a path that crosses his farmyard – and shakes the hives on Sunday mornings. They prove an effective – and so far as I know – a legal deterrent. But the prize must go to the chap who bought a huge hairy Highland heifer and put a ring through her nose before turning her out on pasture where he objected to the litter left by picnickers. Approaching strangers confronted by the huge, long-horned beast, with a ring in 'his' nose, did not look at the other end and, if they had, the long Highland coat could well have deceived them about the fact that there was really nothing to hide beneath 'his' sporran.

Such stupid conflicts would be unnecessary with sensible give-and-take. The Country Landowners' Association did a survey of members from which it emerged that most of them were perfectly happy to welcome strangers on their land provided they were well behaved. The Ramblers' Association, with about the same membership as the Country Landowners', presumably consists of equally rational members but, unfortunately, a strident lunatic fringe let down the majority by refusing any compromise, advocating mass trespass and even flouting the Country Code, accepted by most reasonable people as an admirable guideline for townsfolk and countryfolk alike. The greatest problem with all factions is to prevent the fanatics alienating the opposition so badly that give-and-take becomes a discredited theme because both camps believe that the other will take all and give nowt.

The Wildlife and Countryside Act advocated the use of Sites of Special Scientific Interest to prevent the destruction of important habitats. Until the act was passed SSSIs were declared but there were no teeth to enforce observance. The new act laid down penalties for their destruction, in return for which compensation was offered equivalent to the profit that would be lost by compliance. The weakness was that there was a three-month period for appeal between notification of intent to declare an SSSI and ratification. A few maverick farmers took advantage of this and drained land which was to be an SSSI because of its wetland nature, or ploughed ancient meadows that harboured precious native wild flowers.

The proposal to enforce penalties for altering the site between notification of intent and ratification came not from conservationists but from the Country Landowners' Association, who wished to preserve their public image for responsibility. Protest groups, with no reputation to maintain, would do well to follow suit.

A most stupid clause in the Act has been abused already. It is possible to buy land covered by an SSSI which is of very low value – perhaps because it is very wet. However costly draining it might be, the owner can apply for permission to do so, for conversion to arable land, even though he privately has no intention of going ahead. If the SSSI is important as a habitat for rare flora or fauna, he will be refused permission. He can then claim compensation for loss of the profit he *would* have made! Because there is no ceiling on such claims, there is real danger that they will escalate so much that refusal to grant permission to change the use of SSSIs will be so uneconomic that it will be impossible to preserve them. Add the fact that the Ministry of Agriculture encourages crop production, however uneconomic and whatever treasures are destroyed, and the outlook is bleak until such decisions are passed to more responsible departments.

Access to the extent of keeping to the straight and narrow paths of public rights of way is one thing, but many of those aching with nostalgia for simple country things crave more. They long for the right to go where they like, provided that they cause no damage. It so happens that some of the wildest and most beautiful countryside coincides with the areas where hill farmers barely scrape a living. I use the term advisedly because hill country, best suited to beef and sheep farming, is some of the most difficult terrain on which to earn a decent living. It is impossible to slash costs by mechanising and you can't shed much labour unless you manage with one sheep dog instead of two! The land is too steep for tractors to go far without capsizing and usually too rocky for plough or other implements. Any form of income that can stand in for a cash crop can make all the difference between success and failure – and one possible cash crop is tourism.

Isolated hill farms are now doing bed-and-breakfasts or providing facilities for camping or caravans, but it might also be possible to allow visitors to get their recreation by wandering where they like, and charging for the privilege. Fishermen and shooters take it for granted that they will pay for the rights to enjoy their sport. Yachtsmen expect to pay dues to have their boats on private water and the Forestry Commission charge motorists who wish to run rallies on forest tracks, and campers who wish to camp. Only huntsmen and walkers expect to romp over the land of others for nothing – and the best way of preserving the privilege is to make sure that those who work there and own or rent the land earn enough to make a viable living. Some pursuits are obviously incompatible. It is impossible to do much bird watching if a fleet of powerboats are thrashing around the lake and deerstalkers won't find many deer if there is a motorcycle scramble on the same moor at the same time. Ramblers get very het-up when trail riders, on buzzing motorbikes, come streaming along the same mountain track or drovers' road, so that the sensible solution seems to be some sort of zoning. Large, unlandscaped gravel pits could be set aside for powerboating, leaving more beautiful lakes and pools, which are richer in wildlife, for the quieter pursuits of birdwatching, while stretches of water which lie between the two extremes could be designated for fishing and sailing. So far as is humanly possible, such zoning should be guided by the supply and demand of both owners and participants rather than by the whim of some petty bureaucrat, tucked away in an office, pandering to his delusions of grandeur.

One pastime which has recently mushroomed in popularity is orienteering. Competitors, each with map and compass, are required to find their way as quickly as possible from start to finish across territory where they cannot see where they are going. A great deal of orienteering takes place on land owned by the Forestry Commission and it does not seem to be appreciated what harm to wildlife a crowd of people can do if spread out widely. Shy creatures, like deer, are surprisingly tolerant of small quiet parties of people passing through their territory, even accompanied by a well behaved dog. But if a party splits up, so that some people start to pass on one side of the deer and some on the other, the deer immediately act as if they are evading being surrounded. People on one side leave plenty of scope to escape. But people on both sides at once pose a deadly threat. Imagine, therefore, the effect of several hundred orienteers spread out through the forest

107

like a swarm of hungry ants. Whichever way the deer try to escape, they are bound to be confronted by more strange sweaty men with running shorts and maps. They escape in panic and, at best, stray onto neighbours' untrespassed land, where they may wreak havoc among crops and, at worst, are parted from their newborn fawns. Orienteering should be confined to woods with minimal wildlife potential.

Even naturalists, in theory the quietest and least disturbing candidates for access to the countryside, can cause considerable dissension, especially those who specialise on one aspect till their one-track minds focus narrowly to the exclusion of all else. Some birdwatchers wander round, eyes rolling skyward so that they may not see the wildflowers they tread underfoot. Dyed-in-the-wool botanists wander round examining each bloom and disturbing the birds on their nests overhead. Chaps prancing about with butterfly nets can wreak havoc at all levels and I once had an ancient oak tree ring-barked by an irresponsible university lecturer searching for micro-beetles the size of pinheads, which only live beneath the bark of ancient oaks. I made it crystal clear to him that I was not surprised the micro-beetles, in which he specialised, were rare if louts like him went round destroying their limited habitat.

Odd as it may seem, it is often the specialist naturalists who do most damage. They become so specialised that they know so much about a gnat's knee that they could programme the mechanics on a computer – but they don't know where its ankle is. Archeologists with metal detectors, the curse of modern technology, are destroying irreplaceable sites that have lain hidden for centuries, and moth hunters, with mercury vapour lamps as traps, can drain whole populations of insects, over a wide area, in a night.

Conflicts, large and small, and the whole question of zoning, raise the question of whether planning should be applied to the countryside. Urban planning is an accepted fact of life, though Bath and other beautiful cities were built without either help or interference from any bureaucratic planning office. Nevertheless, the concept of Green Belts, to prevent the increasing sprawl of cities, and restriction of permission to build houses and shops and factories with nothing in mind but profit, have undoubtedly done much to curtail anti-social development despite the occasional discovery of graft in town halls and council chambers.

So far the countryside has escaped the attentions of planners to any great extent, but the grubbing out of woodlands and hedgerows, to make vast prairies of plough, and unsightly concrete and asbestos buildings, with a forest of corn silos, proud as phallic symbols, are so obtrusive that pressure is growing for the application of planning consent for change of use of land or the erection of new buildings. Conservative governments, depending so much on rural votes, are happy enough to lay down guidelines but to leave it to good sense and cooperation to follow them. The snag is that a few of the biggest and richest farmers, including impersonal pension funds and other faceless institutions, continue to do what is most profitable, irrespective of public opinion or official advice or wildlife welfare, getting the rest of the responsible farming community a name that stinks as bad as theirs.

Many socialists would go further and apply planning laws that had to be obeyed, because their votes come from the urban masses, with no stake in the countryside. The country might be more beautiful, as a temporary result – until it went broke. The fact is that farming is a profession where supply and demand change so quickly for reasons such as weather and new techniques, that bureaucratic planning is too cumbersome and too slow to keep pace with sudden fluctuations.

The best compromise would once more be the encouragement of smaller owner and tenant farmers, at the expense of the big boys. Folk who live on the land they work care for it because it is their home and because they know their neighbours personally – and therefore are far more sensitive to public criticism than impersonal pen pushers a hundred miles away. If some of them disliked unannounced visits from unknown strangers, there would still be those who would welcome them as a valued cash crop and be glad to provide facilities for their pleasure.

13

Huntin'

No subject generates more heat than the subject of Blood Sports, as 'antis' regard them, or Field Sports as the advocates prefer. There are two fundamental reasons for objecting to them, the first being that they are cruel and the second that people enjoy them. I regard what other people enjoy as being none of my business, and I have no intention of entering into the philosophical side of *that* argument. I am far more concerned with the cruelty involved and with the effect sport has on the survival of the species or on the welfare of other species.

Otters, for example, became extremely rare (and are still growing rarer) in many parts of the country during the 1940s. The worst-hit areas were rivers which ran through arable country or were heavily used for holiday traffic or coarse fishing. Examination of dead otters – either killed by huntsmen as quarry, or by water bailiffs preserving fishing beats, or by gamekeepers preserving game, or found as the result of road casualties – frequently showed signs of pesticide poisoning. The fashion of controlling agricultural pests by chemical sprays resulted in diabolically dangerous substances being spewed on the land in the 1940s and 1950s. Dieldrin, Aldrin and Heptochlor, three of the worst of the chlorinated hydrocarbons, have since been banned – by consent if not by law – but, for many years, such substances were swilled off the land into ditches by every

111

shower of rain. They not only killed insect pests but frogs, toads and fish, when the ditches drained into streams and rivers and ponds and lakes. Dying frogs and fish are obviously easier to catch than healthy specimens, so a great many herons and kingfishers and otters either died of secondary poisoning or were rendered sterile, if they had only ingested a sub-lethal dose.

Otters are among our shyest mammals and will not put up with much human disturbance. If the banks of a river are dotted with endless chains of coarse fishermen, sitting a few yards apart in competition, the local otters will leave in search of quieter and more peaceful holts. The holts where they live are often beneath the roots of riverside trees, and the modern, objectionable craze of water authorities is to avoid any threat of flooding by converting peaceful rivers into ugly drains, sluicing storm water down to the sea. To do so efficiently, they need mechanical excavators and drag lines, which deepen channels and yank out any tree roots in their way. Such machines, like many who use them, are all brawn and little brain, and they soon cleared the stretches where it really was necessary to improve the unimpeded flow. They are also ruinously expensive, so that the financial gnomes insisted that they were used the maximum number of days a year, to spread their capital cost. As a result, they 'cleared' a great many stretches of bank where it was not really necessary to improve the flow. This obliterated the aesthetic attractions of some of our most beautiful rivers and destroyed the habitat needed by otters to thrive at the same time. It also turned beautiful rivers into drains which sluice precious water to the sea instead of conserving it, so that water authorities are responsible for turning minor dry spells into serious droughts.

The combination of loss of living space, human disturbance by powerboats, canoes and walkers on the banks as well as fishermen, and the poisoning of food supplies posed a great threat to the otter population. Fishing bailiffs and gamekeepers added to the peril by shooting and trapping them. For a while, otter hunters resisted pressure to cease hunting because they rightly pointed out that the population of otters had remained static during the century from the 1840s to the 1940s, as indicated by the number of otters annually found and killed by hounds. If hunting really had been responsible for their decline, they said, it would have been impossible to 'find' steady numbers of otters each year. There is no doubt, to my mind, that it was *not* hunting but a combination of disturbance, bank clearance, poisonous pesticides and trapping which accounted for the slump in otter population.

Some people still dislike them because they have canine teeth, for there are always oafs who will kill any bird with a hooked bill, from eagles to barn owls, and any animal with canine teeth, from pine martens to domestic cats. Any predator is subconsciously regarded as a dangerous competitor. Being creatures of exceptionally fixed habits, otters fall easy victims to such people. Holts, which have been used by generations of otters, will be repopulated almost as soon as the last tenant has stiffened with *rigor mortis*. A farmer, a few miles from where I live, had ground on both sides of a road, beneath which a stream flowed through a culvert. Otters, travelling up and downstream, passed through this culvert. To be more precise, they entered it but never got through because the farmer kept a gin-trap set in the culvert, his human smell masked because the trap was beneath the surface of the shallow water. He boasted that he caught about a score of otters a year. What good it did him is difficult to assess because he neither fished nor hunted – but he did dislike the Master of Otter Hounds. Otters travel for miles up and downstream so that his trap could account for all the otters over a considerable stretch, so perhaps he hoped the Master would keep away if there was no quarry to hunt. Fortunately for otters, he has gone where he can trap no more and, if there is any justice in the after-life, he is awaiting Judgement Day with a gin-trap on his leg as the badge of his misdeeds!

Although otter hunts did not kill enough otters to account for their decline, hunting could well have added the final straw. In any case, there were, to my mind, good reasons for outlawing it on grounds of cruelty. Hounds cannot catch otters on large sheets of water or in deep, wide rivers, because otters can easily outswim hounds – or dive far beneath them – so that hounds cannot force them into water shallow enough to get to terms. Nor could hounds, on their own, account for many otters even in medium-sized rivers if they got no unfair help from human followers. The custom was to line the banks with followers and to form a 'stickle', or human chain, across the nearest shallow water up and downstream from where the hunted otter was known to be. When hounds disturbed their quarry, which dived and swam quietly away, the waiting field watched for the submerged otter to swim quietly by or to disclose its sub-aquatic presence by the chain of air bubbles he exhaled. When he was spotted, hounds were cheered to the spot, and, even if he was still swimming below the surface, the hounds could hunt his scent as the rising bubbles burst. If the field had not given his presence away repeatedly, but stayed behind the hunting hounds, as in other forms of hunting (where 'heading' a fox is bad form) the otter would have escaped nine times out of ten. But the way the sport was conducted, the quarry was continually harried by hounds and men for hours on end until he was forced by fatigue to come ashore or take refuge in a holt. Even there he was not safe because a terrier would be fetched to eject him and force him to run the gauntlet until, at last, hounds caught him.

Whatever price you put on the countryside, the banning of otter hunting did nothing to detract from the value for

114

TOP Non-factory farming does not mean uneconomic farming. These free-range sows and their piglets are competitive with pork produced by intensive methods and their quality is much higher.

ABOVE Traditional farming can involve change to the countryside too. Here old roots from oak trees are being blown up to improve the land.

TOP Hadrian's Wall, where the enormous number of visitors cause erosion problems as people walk along the Wall.

ABOVE Another popular tourist spot, but this one is designed for their needs: the very un-British lions at Stapleford Park, Leicestershire.

OPPOSITE Dovedale, Derbyshire – part of the Peak District National Park.

The author with the Duchess of Devonshire at Chatsworth, which is one of the finest estates in the country.

Successful conservation at work: Shapwick Reserve in Somerset, where old peat diggings have been brought back to their natural state.

me. A few people, notably Philip Wayre, who runs the Otter Trust at Earsham near Bungay, in Suffolk, are doing their best to repair the damage by breeding otters in captivity to restock depopulated rivers where the owners can promise that there will be no further disturbance. But otters are so shy that very exceptional skill is needed to persuade them to breed in captivity and, although the run-off of poisonous pesticides is less than it was ten years ago, River Board officials still vandalise trees along the banks, fishermen, hikers and boaters still turn them into the equivalent of Venice or Piccadilly Circus, and there are still louts who will shoot and trap indiscriminately – whatever the law says.

An unnecessary hazard is the new sport of mink hunting. During the last two decades, a population of wild mink has built up on many rivers started by a few escapes from reputable fur farms. The problem was aggravated by fur farmers who went bankrupt and either did not keep fences in repair or simply abandoned the project. In the last year or so the whole problem has been magnified out of all proportion by the irresponsible lunatic fringe of protest groups who have broken into fur farms and 'liberated' the occupants. The damage mink can cause to our native fauna – and domestic stock – is incalculable and, although Ministry of Agriculture pest officers spend a fortune trying to trap them, they do not employ the calibre of labour with the necessary skill. As a result feral mink now inhabit the banks of many rivers once populated by otters – where there are signs of otters returning due to improved conditions. So packs of otter hounds, denied their traditional sport, have taken to hunting mink instead. They may be conscientious about 'whipping hounds off' when they discover they are hunting otter by mistake, but the disturbance they cause can do almost as much harm as if otters were still their legitimate quarry. It would be far better to line the river banks with cage-and-tunnel traps, too small for otters to enter, because mink are such curious creatures that it is far easier for skilled men, if not pest officers, to trap them than it is to catch them with hounds.

No other sport with hounds provides so many reasons to prevent it. Anyone who examines a one-inch Ordnance map for most parts of the country cannot fail to appreciate how much of our aesthetic pleasure is derived from the glorious patchwork of farms and small woodlands that is so typical of the English countryside. The woods were not spattered there by random chance. They were planted in strategic spots deliberately and, as the names of so many of them testify, they were originally planted as fox covers.

During the breeding season, roughly from March to the end of May or beginning of June, vixens have cubs in underground fox-earths and the cubs remain there, being fed by the dam, till they are strong enough to begin hunting food for themselves and fleet enough of foot to escape their enemies. Vixens sometimes cub in large setts that were originally dug out by badgers, occasionally using a sett that is still the home of Brock. Sometimes the vixen opens up a rabbit warren – which may also still be shared by the original occupants – and sometimes she will cub in an artificial earth constructed for the purpose by the servants of the hunt. These artificial earths – or 'drains' – are of very simple design, consisting of a den about two feet by three feet by a foot deep, buried near the surface of a secluded bank. The sides of the den are usually timber – for warmth – but may be sandstone or occasionally brick. Two entrance tunnels, usually drainpipes of nine or ten inches in diameter, several yards long in a horseshoe curve (to avoid wind blowing direct into the den) slope down from the den to give entrances lower down the bank. The whole is covered with earth and turfed over, to blend with the secluded surroundings, and it is as likely to be adopted by foxes as an artificial nest-box, cunningly sited, is to be filled with a brood of blue tits. The construction of these artificial earths is criticised by opponents of fox hunting, who claim that foxes are 'bred' and 'turned down', or liberated, to be hunted. The fact is that they are adopted by truly wild foxes and only tenanted permanently while vixens are breeding, when hunting is out of season.

Out of the breeding season, foxes spent most of their time above ground in thick cover, where they are unlikely to be disturbed. When cubs leave the breeding earth – frequently driven out by the density of fleas and sheep ticks – they often lie-up in standing hay or corn. When that is harvested they move to root crops or standing kale and, when that goes in winter, they like to lie in the dense undergrowth of woodlands. Woodlands which are large and extensive harbour foxes well, but it is very difficult for hounds to drive them out as their instinct tells they are far safer dodging about in thick cover than they could be if they took to their heels across open country.

Foxhunters coped with this by planting small woods, often with rhododendron understorey, specially to encourage foxes to lie-up there. An added attraction, in the shape of the artificial earth, or drain, was often added to encourage breeding in fox covers where their safety could be supervised. Foxes also came to rely on the sanctuary of their drain when put under pressure by hounds. Any confidence they had in such spurious safety was misplaced because an earth-stopper was sent round, the night before hounds met, to run his terrier through all the drains, making certain they were untenanted, and to 'stop' them with a bundle of brushwood or sod of turf so that any fox, seeking safety there next day, would find his entrance barred and be forced to 'point his mask' across open country, at what he believed (usually wrongly) would be safe sanctuary in another earth.

In order to get exciting runs, over open country, it was only necessary to plant woods and spinneys at strategic points so that, as soon as hounds were put into cover, they would find a fox, which could not retreat into the already 'stopped' earth so would naturally take flight in the direction of another familiar hiding place in a cover, anything from a few fields to a mile or so away. This would also be barred, so on he would go and, the more cleverly the covers were sited, the more exciting the sport that hounds would give.

The fringe benefit, for those not involved with hunting, was – and is! – that the whole countryside in hunting country is dappled with little woods and spinneys, osier beds and coppices, which merge with the patchwork of arable and pasture fields to give a diversity of shapes and colours that is unique to the British countryside. Pure economics could never justify the space they occupy. Huge acreages of woodland are owned either by landowners, who enjoy hunting, or by the hunts themselves. They are vitally necessary for the continuance of sport because no quarry will run across open country unless it knows that ahead is what past experience has shown to be safe sanctuary. The fox, of course, has no means of knowing it may find the earth 'stopped' when it gets there. If it did not believe it was heading for safety it would simply dodge about, in efforts to elude hounds, in the nearest cover it could find, from a field of kale to a housing estate, with the inevitable result that the sportsmen would not get the exciting chase they sought.

It follows therefore, that if foxhunting was stopped for any reason, it would be far more profitable to fell existing covers and plant them with crops than to leave them as they are. However satisfactory those opposed to sport might find prohibition of fox hunting to be, the inevitable side-effect would be a rapid deterioration in aesthetic rural scenes, and very large numbers of countryfolk, employed in trades connected with hunting, would be made redundant.

Whether abolition is desirable on the grounds of cruelty

to foxes is a matter of subjective opinion. Whatever the lunatic fringe of fox hunters say about 'foxes enjoying being hunted' only reinforces the view of those who believe that too much time spent in the saddle shakes men's – and women's – brains down into their breeches. Casual observations of the sexual habits of horsey folk confirm the impression that, apart from hunting, many have but one thought in mind! The plain truth is that foxes do *not* enjoy being hunted, still less being caught, although death at the fangs of a pack of hounds is probably far swifter than opponents pretend. The inescapable fact is that most alternative ways that foxes perish are far worse. I live in an area where very large tracts of woodland make hunting too difficult to be very effective. It does not pose much threat to the local foxes. As a result there are a lot of foxes – and every man's hand is against them. Although very few people keep free-range poultry – and poultry cooped up in intensive units are safe from attack – foxes are snared because they always were snared when vulnerable poultry ranged the fields. Keepers snare them legally and poison them illegally, while their masters blast them with shotguns if beaters drive them out while pheasant shooting. Vixens are dug out with terriers when they go to earth to cub – and the cubs are hit on the head or left to starve, as luck will have it. My experience is that the way most foxes meet their end is no more pleasant than being overwhelmed by hounds.

As a result, I am neither personally for nor against fox hunting, since I do not set myself up in judgement of what people should or should not enjoy and I see no prospect of foxes being allowed to proliferate if hunting ceased. But it so happens that I do not think hunting has a long future for a number of reasons. Hunt saboteurs are unlikely to have much effect because they are precisely the type of mindless louts who endanger our wildlife by letting out mink or who swell the left-wing mobs indulging in any sort of 'protest'. Sensible people are fed up with such militants and society will have to bring them to heel or suffer disintegration. Modern arable farming is far more likely to sound the knell for those who wish to canter over the countryside.

In the old days before the agri-revolution, it took so long to harvest corn, stook it to ripen and cart it to the rickyard that there was not time to plough the stubble before the autumn weather broke. So every arable farm had field after field of stubble, interspersed with grass fields left because it was the custom to rest ground by rotating crops from grass to corn to roots. Nobody minded huntsmen galloping over such land for the simple reason that they did no damage when they did.

The type of men who hunted in those days knew the significance of the short shoots of winter sown corn, just sprouting, and were far too well mannered to put a horse's hoof on it. Half the jokers who go out now would think it was turf and gallop on oblivious. Even where hedgerows are allowed to exist, labour costs too much to cut-and-lay them so that they are roughly slashed with mechanical hedge-cutters, the gaps being plugged with barbed wire, while many fields are surrounded by wire-mesh fences, topped with barbed. No place for any horse to jump. At the moment, hunts have compromised by having Fieldmasters, who are responsible for marshalling the followers, so that they do not go onto vulnerable crops, and they even have to have gate-shutters, because the modern school cannot jump and has not the manners to close gates behind them. The writing is on the wall and, whatever one thinks about the ethics of hunting, the prospects for its future are not bright and, when it goes, steps will have to be taken to preserve the scattered woodland associated with it – or the countryside and other wildlife will lose a lot.

There are, of course, other quarry but fox and other forms

of hunting. Just how subjective the objections to them are is illustrated by the views one hears on stag hunting and ratting, for example. 'Stags are noble beasts,' the antis say, 'and it is wicked to hunt them.' Try pressing the same chap about his views on ratting and the chances are that he will believe rats are 'dirty' pests and nothing is too bad for them. Looked at objectively, it is not difficult to believe that terror and pain, to a rat, are probably identical to terror and pain to a noble stag! And it is certain that the red deer on Exmoor are only tolerated because the farmers there like stag hunting. They would be shot the first season that hounds were taken away because of the damage they do to crops – and the market price of venison.

Beagling – or hunting hares with short-legged hounds – is demonstrably more cruel than coursing hares with greyhounds, because greyhounds are sprinters and either catch, and kill, their quarry within about two minutes – or the quarry outruns them and gets clean away. But it is the only hound sport in which, because of the brevity of the chase, spectators can see the whole drama, from the time a hare, native of the local fields, is driven past the slipper, with two hounds, to the conclusion of capture or escape. Beagles, on the other hand, start a hare from the concealment of her cover, when she immediately disappears over the horizon. The little hounds then hunt her by scent, coming up to her and starting her from her form each time she hides in cover, until at last they literally wear her down so that she cannot even outpace her sluggish opponents whose only asset is that they can keep going forever. It is a ruthless war of lung-shattering attrition, but opposition is minimal, partly because the antis don't understand the significance and partly because, being so left-wing, they can forgive all if they can persuade themselves it is a 'working men's' sport, where you don't have to dress up or ride on expensive horses, as the upper classes do!

What Price the Countryside?

Hunting, in one form or another, has been so closely involved with the countryside for centuries that the very appearance of the countryside owes an enormous amount to its influence. The alternative means of disposing of most quarry species leave much to be desired but inevitable trends seem likely to limit the future of most conventional hunts. Whether you mourn or gloat, when that time comes is a matter of subjective opinion, for there is much to be said on both sides.

14

Shootin'

Like it or not, whatever price you put upon the countryside cannot be divorced from the sport of shooting. Just as the covers planted to harbour foxes have had an immense – and beneficial – effect upon the rural scene, so have the covers planted to harbour game, especially pheasants. On some estates, keepers are expected to rear pheasants *and* harbour foxes in the same covers so that, when the hunt comes, they can be certain of finding a fox. A great deal of play is made of the good relations between hunting and shooting men, which is reasonable enough since both are subject to attacks by the same opposition. Add to that the fact that there are still a few rich men who can afford to hunt twice a week and shoot three times, and it is obvious that they will expect to be shown sport in whichever sphere they happen to engage.

In my young days, the only folk who could afford such sport were the great landowners, who still owned estates large enough to find sanctuary for quarry that either ran or flew. Nowadays, shooting has become so expensive that even the largest landowners let off much of their land to sporting syndicates of brassy tycoons, who can afford to shell out several hundred pounds for each day's shooting and, because so many of them are spivs, with more money than manners, their influence on the countryside is often catastrophic. Many of them, who made their fortunes late in life, couldn't ride a clothes-horse, far less a hunter, so

123

that their common ground with the hunting fraternity is minimal. Because money means so much to them, they are utterly intolerant of anything that threatens their precious pheasants. Their prime aim is to kill enough pheasants to recoup the costs of the shoot from the sale of the game their 'sport' provides.

As a result, the first quality they demand in their game-keepers is to be 'good with vermin'. They ask no questions about how vermin is caught – and their keepers wouldn't tell them if they did. Snares, traps, the gun and poison, legal and illegal, are all employed and all forms of vermin, including foxes and the neighbours' cats get a one-way ticket if they trespass on the shoot.

But it is no use having such a status symbol unless you are liked and respected by your peers – and the only peers of such Flash Harrys are a minority of similar types from the hunting field. It is mutually understood that hounds should not disturb their coverts until the shooting season ends on February 1st – or after the Christmas shoot at the earliest. But, when they do come, it is essential to show them good sport by letting them find a fox, which will prove, by implication, how cooperative they are. This may not be easy, since every fox stupid enough to poke his mask over the shoot boundary, since hounds dispersed cubs in autumn, has probably been quietly knocked-off and stuffed, brush first, down an inconspicuous rabbit hole. One solution is to dig out a fox on neighbouring land and turn him into cover on one side, as hounds are cheered in on the other. It is, of course, strictly illegal to turn down, or release, such 'bagmen', so the keeper does not mention the fact to the boss, who is doubly impressed that he has found a treasure who can show pheasants in the shooting season – but find (?!) a fox in the same cover when his mates in huntin' pink arrive.

In the old days, a man who owned, or even rented, a shoot would have been grossly insulted if anyone had offered money for the privilege of being a guest. Shooting was reserved for personal friends – and it gave equal pleasure to invite them for a day's shooting and to accept, in due course, a return invitation. Such shooting was more leisurely, in terms of time, than present shooting parties, but far more energetic. Small parties of five or six guns would set off, with a keeper and two or three beaters. They would all string out in a line, a beater between each pair of guns, so that they were not in each other's pockets, with consequent risk of poaching each other's birds. No guest took his dog unless he had asked permission from his host or had been specifically invited. But shooting men in those days had more time to spend on handling dogs and many of them owned animals which would not have disgraced themselves in modern Field Trials, though they were expected to show more initiative than modern competition dogs and not rely on hand signals from their boss.

The whole line, guns, beaters and dogs moved forward to flush their own game before shooting it. During a day, such a party walked miles through fields of roots and kale, through woodland and reed beds, osier plantations and thick cover. Their dogs left no bramble patch nor furze cover untried and, apart from a frugal picnic of beer and sandwiches, they slogged on till dusk was falling. Bags were often quite small, because few estates reared birds intensively, only a few hundred pheasants being hatched and reared by broody hens where as many thousands are reared today with incubators and bottle-gas brooders. Quantity was not the prime object then, as the game registers recorded generations ago confirm. Such old-fashioned sportsmen liked 'sporting' shots and they would rather allow an immature or otherwise easy bird to fly away unscathed than pull their trigger on anything approaching a sitting duck. On only a few big days did they have a series of drives for driven birds.

Most wild creatures are of very fixed habits, having

favourite feeding places and favourite sanctuaries. Keepers know and exploit this, so they rear their pheasants near one cover – usually woodland – and gradually entice them to feed in another. Ideally these two woods should be a quarter to half a mile apart, on the highest ground in the area. When the time comes for the shoot, the guns are placed in a line on the low ground between the two covers so that, when beaters enter one cover, to flush the birds, they seek sanctuary by flying to the other wood, passing high over the waiting guns below. By the time they pass over them, they are flying very fast and, if there is a decent wind, they will also be flying in a curve, which makes them very hard to hit, so that a high proportion reach their haven unscathed. Unfortunately, because they are so far above the guns, some of them will not be killed outright but will reach their destination wounded. 'Pricked' birds, shooting men term them, which must be the most callous understatement of the season. If the men were 'pricked' as badly, they would be in intensive care. It is therefore absolutely vital to have efficient dogs, to seek out wounded birds and bring them for humane dispatch, and the more leisurely parties of the old days had plenty of time and took great pride in making sure that every bird they had hit had been safely retrieved before moving on to the next drive – however long it took.

Their modern counterparts are often not so fussy. The maximum number of saleable birds is their object and they won't waste time chasing problematic pricked birds, which their inferior dogs may not find, if they could have been more profitably employed at the next drive. They write them off as the acceptable losses of their sport. They don't mind wasting time at lunch, though. This is usually held at a local pub or a keeper's wife will provide a hot, well-cooked meal in her best room. This will be a real expense-account affair, which means that no expense will be spared because it can be written-off against tax on the books, usually of some company run by a member of the shoot. Wine and spirits flow like water so that, when sport resumes some time in the afternoon, no insurance broker in his right mind would insure either guns or beaters against the risk of ending up among the pricked birds. The tax fiddles of such shoots are legion. Some claim they could get no export orders without inviting foreign customers to shoot with them, so that the expenses of running the shoot should be met by the taxman, and a foreign customer is invited to justify the claim. Keepers are sometimes on factory payrolls, probably entered in the ledger under the heading of security. Vehicles used for the shoot often appear against some factory transport department.

Such Rolls-Royce-and-Runny-Nose tycoons do not even flatter themselves with the title 'sportsmen', for they refer to themselves as shooters and, if anything moves, they shoot it. Pheasants which survive long enough to get properly airborne are fairly lucky because numbers in the bag, not difficult shots, are the criterion of excellence. But, sportsmen or shooters, the modern school or old-fashioned sticklers for etiquette, the effect they have on the quality of the countryside is still prodigious.

Game cannot survive 'wild' in the countryside without suitable habitat. It needs cover for concealment and food plants for sustenance, and both shooters and sportsmen need covers from which to flush their quarry and more covers for the survivors to seek sanctuary. Shooting rents of several pounds an acre (more than agricultural rents when I was young!) ensure that, even when farmers are not themselves interested in shooting, their shooting rights have great potential, as a cash crop – but only if they maintain suitable habitat. As a result, farmers who enlarge fields to suit the needs of modern mechanisation are likely to grub out no more hedges than necessary, because hedgerows are good breeding cover for pheasants and partridges, and pheasants will squat in hedgerows too. By the same token, the small woods and covers, often on unploughable hill tops, which were planted to improve our ancestors' sport, are valuable assets when fixing shooting rents.

Small pools, left in field corners as drinking places for farm stock, will attract good stocks of wild duck if grain is scattered there to feed them. An hour's 'flighting', concealed round such pools at dusk, when conventional shooting is over, provides an added bonus, as duck fly in to feed, for which shooting tenants are happy to fork out more rent. So hedges and pools, cover in odd field corners and picturesque small woods are all valuable assets to any farm where there are opportunities to cash-in on demand for sporting rights, so long as it persists. Without such demand, such odd corners would be an impediment to tidy farming, so that they would be bulldozed flat to add a few more acres of grain or grazing.

Cover which is suitable for pheasants and other game is also attractive to a wide range of wildlife – provided the gamekeeper allows it to survive. In the old days, when sportsmen treated game with great respect, they were often most destructive to other forms of wildlife. The manners and ethics of old-fashioned sportsmen were, in most respects, impeccable. The one exception was the treatment of predators as enemies of game. Every keeper's gibbet was decorated with a grisly array of rotting corpses, including crows and magpies, jays and hawks and owls, stoats and weasels, badgers and hedgehogs, rats and the tails of poaching cats. By the keeper's definition, *any* cat found on his beat would not be there if it was not poaching – and it would not return home if he could help it. The only reason that fox brushes did not figure among the trophies was that killing a fox, except with hounds, was vulpicide, a socially unacceptable crime, for which the keeper's master would have been ostracised by all respectable people! So foxes which trespassed were not displayed but put discreetly out of sight.

It is easy enough to understand the reasons for such ruthless action against predators. Intensive husbandry was still a blessing – or a curse! – which had not yet burst upon the countryside. A high proportion of pheasants and other game (including partridge) were hatched and reared in the wild with no artificial aid but the provision of suitable habitat, which happens to be very beautiful. If predators such as carrion crow and magpies, stoats and wandering cats had the freedom of the same habitat, the predators would thrive and the game would be obliterated. So keepers trapped and shot the predators to remove the threat to wild game. At this time large numbers of pheasants were hand reared by putting the eggs under broody domestic hens to hatch. The hens and chicks were then put out in row after row of coops, which imprisoned the hen but allowed the chicks to forage, within earshot, round the coops. If a predator attacked the chicks, the hen was powerless to help because she couldn't get out of her covered coop, and, if the hens had been allowed to range with the chicks, they would have fought among themselves and killed each other's chicks. Any chance of rearing a successful crop of game therefore relied on close control of predators.

Things are different now. Very few truly wild pheasants are shot. Hen pheasants are caught up in cage traps at the end of the season and kept in covered enclosures, or laying pens, where their eggs are collected daily to be hatched by the thousand in electric – vermin-proof! – incubators, or hatching eggs or day-old chicks are purchased from commercial game farms. The chicks are then reared in bottle-gas or electrically heated brooders with covered, vermin-proof, runs attached for them to exercise. At about three months old, when they are too large for most predators but foxes to worry, they are put out into large release pens in the wood. These pens are surrounded with netting six feet high, around which is a low strand of electric fence, to prevent foxes digging under the netting. The poults are fed in these release pens and they gradually learn to fly out to forage in the surrounding wood. They have not the sense to fly back to feed at evening but walk home and pace up and down the fence, looking for a way in. The keeper provides this by making holes, at intervals, in the bottom of the fence, large enough for young pheasants but not foxes to creep through. To prevent them creeping out again, each hole has a funnel of inward-pointing netting through which the poults creep. They are not bright enough to find the exit to creep out again, but pace inside the netting till roosting time. Next morning, when they are roosting in trees near the fence, it is easy to fly over the top to forage.

This system of husbandry is not nearly as labour-intensive as the old coop and broody system, so far less keepers are kept. Since the threat of predation has been largely removed because poults are not released in the wood till too large for most predators, modern keepers need to be more skilled in intensive poultry husbandry than in trapping hawks and owls – and fewer keepers means they have less time for such pursuits. So it is only the occasional Bob-the-Killer keeper, beloved by the worst type of shooting syndicate, who still gets his fellows a reputation as foul as his own. The result is that there are still far more small woods and pools and spinneys in the countryside than there would be if shooting was out of fashion, while more predators survive than if the woods were cleared and the countryside became a monoculture of corn or oilseed rape.

The other forms of sport which have great influence on the appearance of the countryside are grouse shooting and deer stalking. Grouse are game birds which require wild moorland for their habitat. They are far more difficult to rear intensively than pheasants and so demanding in their diet that they will only thrive where there is young heather.

The only practical way to increase their numbers is by highly specialised management of their habitat so that the moor consists of a sort of draughtboard pattern of heather of different ages, long thick heather for nesting and short, tender shoots to eat. This is accomplished by burning off the moor in carefully controlled patches, burning the oldest heather which then regenerates so that the grouse can eat the young shoots.

So far as predators go, foxes are the most dangerous, though hen harriers can do some damage and eagles, hunting the skies above, can cause grouse to migrate to safer parts. Eagles and harriers are so scarce that no responsible sportsman would harm them, apart from the fact that they are protected by law. In point of fact, laws are of little use in such circumstances because the habitat is frequently so secluded and wild that the chances of being found out are minimal. My wife and I were invited out to dinner, a year or so ago, on August 12th, the first day of the grouse-shooting season, and the meal was late because one guest had not arrived. He was a bumptious merchant banker, one of the most money-conscious men it has been my misfortune to meet, and when he arrived he made a beeline for me. He opened the conversation by boasting how many brace of grouse they'd shot, mentioning that he'd accounted for two hen harriers. 'I suppose you would not approve?' he asked me. As a guest in someone else's house, it was not possible to tell him what a lout he was, but I understand I made my feelings plain enough for him to get the message and for more responsible guests to signify agreement.

The most important fact about grouse shooting, so far as the scenic value of wild country goes, is that it is one of the main defences against demands to reclaim the wild moorland for agriculture, by clearing and seeding down to pasture, or turning it into vast forestry plantations. There is, of course, violent opposition to either of these alternatives by environmentalists, who love the scenery of remote countryside, but apart from a cacophony of strident squeaks they carry little weight if the economics of forestry or farming can be shown to be vastly superior to the sporting values. Unlike the piners for pretty scenery, devotees of grouse shooting are prepared to put their money where their mouths are so that the value of a moor is directly proportional to the number of grouse it is reasonable to expect to shoot there.

With proper management, moors can support a population of both grouse and sheep. This is vitally important because grouse populations are prone to violent fluctuations for a number of causes, mainly connected with susceptibility to disease and, when grouse are thin on the ground, the extra income from sheep is obviously welcome. One snag is that ticks, carried by sheep, are associated with louping-ill which can decimate grouse stocks, so it is essential not to overstock with sheep. An even worse danger, when sheep are overstocked, is that overgrazing causes decline of the heather. It is replaced by rough, feggy grass which is useless to grouse and unpalatable to sheep. Matters are made worse when sheep are kept on the moor in winter by giving them subsidiary feeds of hay and providing feed blocks, designed to stimulate the ingestion of more and more fibre. The sheep, instead of spreading evenly over the moor, congregate round feed points and graze the heather there to extinction, gradually extending the decline to the rest of the moor. The last resort is forestry, which is the most unpopular choice of all with conservationists and environmentalists alike.

Rich grouse shooters, such as my fellow guest at the dinner party, are prepared to buy or rent grouse moors at very high prices, provided always that they will have opportunities to shoot large numbers of grouse. To this end, it is vital that the moor is covered by healthy heather, which can only be kept in good heart by skilled rotational burning, to ensure there are always patches of tender shoots to feed the grouse. By coincidence it is precisely what conservationists want, so it is vital that they can both shake down in the same bed, however divergent their interests may be.

The other sport that has a major effect on the prospects of moorland, especially the Highlands of Scotland, remaining viable is deer stalking.

Contrary to popular belief, Red Deer are really woodland animals but they do so much damage to commercial woodland, by stripping the bark from trees with their teeth, and fraying small trees with their antlers till they are ripped to ribbons, that they have been driven onto the open uplands.

Their natural breeding rate is such that about fifteen per cent of their population has to be culled or otherwise removed annually to maintain a steady population. Their instinct for self-preservation is sharp enough to make them very difficult to approach as soon as they grasp the fact that men are likely to be armed, so that the most effective way of controlling them is to shoot them with rifles which are lethal at long range. Even so, it is a difficult procedure involving considerable woodcraft, so that the bidding is keen for the sporting rights of any well-stocked moor. As with grouse and salmon, the rents obtainable will have little to do with acreage or yards of river bank. They will depend, instead, on the potential 'bag' obtainable, which will be calculated by taking the average of the number taken annually over a number of previous years. Several hundred pounds a potential stag in the bag will be offered, which is likely to be far more than the rent for sheep and higher still than what forestry operators would offer.

Unfortunately deer and sheep suffer from the same species of worm and other internal parasites, so that there is frequently animosity between local crofters and stalking interests, because the stalking will deteriorate if the sheep population is large enough to pose danger to the deer, quite apart from the fact that overgrazing by sheep obviously limits the potential size of the deer herd.

Without the economic advantages that deer bring to open moorland, in terms of rent obtainable, they would quickly be shot and poached to extinction, by less humane methods than are conventional for skilled marksmen with rifles. Certainly without either deer or grouse there would be grave danger that moors would first be overgrazed by sheep and then turned over for forestry, when useless for other interests. The prospects of other amenity users, such as walkers and campers, forking out enough for the privilege to compete with forestry, far less grouse shooting or stalking, are minimal.

But deer are one of the few species that are increasing nationally in other woodland areas. Fallow, roe and muntjac deer have all multiplied, since the war, to the point where foresters suffer enough damage to treat them as pests. The Forestry Commission, however, takes a far more responsible stance than would be expected from other bureaucratic bodies. Instead of killing as many as they can, they treat the problem scientifically and *tolerate* as many as possible short of unacceptable damage.

Their policy is to survey the forest to estimate how many deer the available keep is capable of sustaining without hunger driving them to do unacceptable damage. The total number of deer present are then counted, usually during the October rut, when bucks collect the does together and it is relatively easy to get an accurate count. When the number of deer present is more than the forest can carry, predation on adjacent farmland is avoided by culling the surplus.

This culling is done, selectively, by skilled marksmen who select the *worst* specimens, leaving the best to breed and improve the quality of the survivors. The reverse often happens where conventional stalking is let to rich foreigners, who naturally want to take the finest heads home as trophies, and are prepared to pay far more than the price of the venison for the privilege.

The Forestry Commission game wardens earn more than their wages from the Commission from the venison taken from their culled deer. An additional sum is collected by allowing licensed shooters to shoot the deer but, in this case, the shooter is accompanied by a warden, who ensures he does not take the deer wanted for breeding stock. Licensed shooters on Forestry Commission land also have to satisfy the Commission that they are expert marksmen by passing a test before they are allowed to shoot a deer. It is a pity that 'sporting' stalkers on private land do not have to do the same.

Apart from satisfactory diet wildlife needs sensible seclusion. Public access to the whole moor, especially in the breeding season, will inevitably mean a poor crop of grouse – which will start the chain reaction from decline in value to conversion to a forest. It should not be impossible to convince any sensible members of the public of this simple fact, but not all members are sensible. The left-wing fringe object to sportsmen enjoying the privilege of being on open moorland, even on shooting days, despite the fact that they are willing to stand the cost of maintaining the heather, which prevents destruction of the moor by forestry or farming. And greedy fools who destroy marginal, protected predators like harriers simply make the prospects worse. At the other end of the scale, militants, including hunt saboteurs and ramblers, actually advocate public trespass where grouse are being shot to make the sport impossible – which would have the inevitable result of destroying the very things they are trying to protect. As with so many other conflicts in the countryside, it is difficult to persuade such extremes to give as well as take.

15

Fishin'

There are more than three and a half million anglers in England and Wales alone, which is more than the number of football fans. The Labour Party includes a ban on hunting in its manifesto, presumably because it believes it to be a minority rich man's sport, in spite of the fact that at least one pack of hounds is owned and hunted by coal miners and many other workers, from pop stars to politicians take part. The same party does not promise to ban shooting or fishing, presumably because they are considered to be more sports of the working class – perhaps because the vast number of participants are directly proportional to the possible loss of votes by alienating them. The question of cruelty does not arise!

I mentioned earlier that my views on field sports have nothing to do with the question of whether people take pleasure from either hunting or shooting or fishing, which I regard as a personal decision which is none of my business. All sport involves some degree of cruelty so that it is relevant to examine the alternatives for the quarry involved. In the case of fox hunting, I believe that snaring and shooting (except at short range by expert marksmen), poisoning and gassing probably involve worse suffering. Shooting involves pricked birds left to die painfully.

But Labour wouldn't ban it. I mention the fact in passing to show how often logic flies out of the window when such emotive subjects arise.

Fishin'

Because fish are cold-blooded, there are many people who conveniently believe that fish do not feel pain. Recent research has shown conclusively that they do, so that philosophical arguments about catching fish for sport should not really differentiate between hunting and shooting and fishing – but they do. I heard a recent broadcast by a noble bishop in a wireless programme on Sunday, in which he said that he would ban fox hunting, 'because it was cruel', but that fishing was all right, because he ate the fish. Thinking that this was a typical case of the Church splitting ecclesiastical hairs, I wrote to ask him whether he believed fish felt pain and whether he objected to the exciting sport of ratting. I did not bother to point out that the reason he had no conscience, because he ate the fish, might well be that, as a noble bishop, he could afford the upmarket sports of salmon and trout fishing, while the poorer members of his flock were limited to coarse fishing, where the quarry is not eaten but put back, to be caught again. And again and again, feeling pain each time! I need not have bothered to develop the argument because his secretary wrote a brief note saying the bishop was too busy to reply!

The incident illustrates the hypocrisy and double standards that colour the arguments about the emotive subject of Field Sports or Blood Sports, depending on which side of the fence you are on – or even if you are astride it, as the bishop was. I can see both sides and I believe that most sports, with the possible exception of foxhunting, will continue for the forseeable future, whether we like it or not, so that it is more profitable to examine the effect they may have on the species or the countryside.

There is no doubt that the huge number of active fishermen gives the Anglers Cooperative Association, their official mouthpiece, a very powerful rural voice. The ACA has done immense service to the countryside by applying great pressure to alleviate trouble caused by pollution, so that some of our rivers are now purer than at any time since the first Industrial Revolution. During the last two or three decades, there has been a tremendous increase in the use of chemical weedkillers and pesticides. Even when correctly applied, a sudden thunderstorm can swill the poison into ditches, from which it will drain into streams and rivers, killing plants and fish. This is an all too obvious hazard, which some farmers, especially the large, efficient, absentee Big Business type tend to shrug off as unavoidable. Not so the ACA, and the efforts they have made both to get laws tightened and enforced have had excellent effects.

There are, however, less obvious but equally dangerous hazards, including the liquor that runs off from silage clamps. Silage is simply the modern alternative to hay. To make hay, the grass is cut and left to dry in the sun and wind, turning it over as necessary to dry the underside. 'Making hay while the sun shines', is all very well except that in our changeable climate the sun does not always oblige at the right time, so that the crop all too often finished up as mouldy mush instead of crisp, sweet-smelling hay. Silage does not have to be dry, it only has to wilt. It is then collected and put in pits or raised clamps and pressure is applied to squash it tightly together, usually by driving a heavy tractor backwards and forwards over the clamp. Damp grass, under pressure, generates heat, and the whole clamp becomes a palatable and nutritious feed instead of dry hay. Wilting, which takes from four to twenty-four hours in dry weather, is the most vital part of the process. In wet weather, there is a great temptation to take a risk and carry the grass to the clamp, as there would be with hay. But wet hay will only go fusty while silage will exude a liquor which is four times as powerful as pig slurry and one 400-tonne clamp will exude as much poisonous matter as the untreated sewage from a town of 150,000, or twice the size of Gloucester. If this seeps into a river, the havoc it will cause to fish and plants is obvious

– and fishermen do not stand meekly by and see their quarry killed. So they let the world know through their mouthpiece, the ACA, which includes a highly specialised legal department, capable of dealing with the National Farmers Union, Ministry of Agriculture or local authorities when harm to fishing interests is threatened.

There are two distinct types of fishing, coarse and game. The quarry of game fishermen are migratory salmon and sea trout, brown trout which may be truly wild or bred in hatcheries, and rainbow trout, which are almost always bred intensively and put into lakes or streams. The most exclusive type of fishing is for salmon, which are caught on artificial flies, often in fast-flowing rivers amongst remote, rugged and beautiful scenery. Salmon are migratory fish, which spend a long time at sea but come back to the river of their birth when the moment comes to spawn. They travel sometimes hundreds of miles upstream, till they reach the shallow gravel beds, bathed by turbulent water, where instinct impels them to spawn. While they are 'clean-run' from the sea, they are caught on artificial flies by fishermen who will fork out a fortune for the privilege, so that a yard of river bank can fetch more than acres of land behind it. The fish are wary and great skill is needed to outwit them but, when caught, their flesh gives pleasure to bishops and barons and anyone else who can afford it. When so much money and privilege is involved, it is obvious that river owners will go to any lengths to preserve their amenity so that there is no shadow of doubt that salmon fishermen make invaluable contributions to the conservation of rivers where salmon survive. Sadly, with so much at stake in areas so remote that law-enforcement is impractical, some of the water bailiffs employed to nurture the salmon are not above destroying any predators, from otters to rare birds.

Second only to salmon fishing, fishing for truly wild trout, particularly in gin-clear chalk streams like the Test and Itchen, is a pastime for which people are willing to dig deep into their pockets. The fish are caught on artificial flies, and great skill is needed to choose the correct fly, usually resembling the insect which has recently been locally hatched, on which the trout are feeding. The clarity of the water magnifies every error the fisherman makes, so that he must not only have the right fly, but attach it to a line so thin that it is practically invisible and 'cast' it with such delicacy that it resembles a wild insect settling on the surface of the water. The purists of the sport do not disturb the river banks by overcrowding and they are ruthless in their attacks on any who pollute the water. They attack fish farmers upstream, who spread diseased waste from their intensively farmed fisheries, and farmers who kill fish with agricultural chemicals or overstimulate plant growth with run-off from artificial fertilisers.

Coarse fishermen come further down the social scale. They are the chaps who sit a few yards apart on crates of beer along the banks of canals and sluggish rivers, large pools and gravel pits. Rain or sunshine makes no difference, for they shelter or shade themselves under huge umbrellas, fishing for roach or perch, bream or dace or almost any of almost forty species of fish which live in our fresh water. The trout or salmon fisherman has to watch the rising fish, to assess what species of fly is currently tempting them, select a similar artificial fly – camouflaging his hook – and then cast so that it lands 'naturally' on the water. He either casts it upstream of the rising fish and lets it float down on the current, if that is what the natural fly are doing, or drops it over the nose of the fish, to tempt him to rise and take it almost as it falls. Once hooked, he has to play the fish until it is exhausted and he can net or gaff it when he brings it to the side. His tackle is so light that undue strain will break it and his quarry will escape. Stronger tackle would be more conspicuous and put off shy and cunning fish. Compared to such finesse, the coarse fisherman's craft

seems dull labourer's work. He only has a few yards
between him and the next competitor – for coarse fishing
is a highly competitive sport – so that the 'swim' where he
can fish is somewhat limited. He baits the bottom of the
swim with maggots or bread or dough or grain or a hun-
dred-and-one secret recipes, to persuade as many fish as
possible to congregate near his swim. He then baits his
hook, often with a worm or blowfly maggot, and adjusts
the depth to which it sinks by adjusting a float on his line.
Then he settles down to watch the float which bobs up
and down when anything moves the bait beneath and bobs
right down when a fish swallows it. Superb coordination
of hand and eye are vital at this stage because he must lift
his rod and 'strike' at the precise moment the hook is in
the fish's mouth. Left a second too long is all that it takes
for the fish to be suspicious and spit out bait and hook and
all. Too early a strike snatches the hook away before the
fish has taken it. The happy medium hooks the fish, which
can be quickly landed because the tackle is strong enough
to yank all but monsters onto the bank.

Coarse fish, living in muddy water, taste pretty foul, so
they are unhooked (or 'disgorged') and kept alive in a
submerged keep net until the end of the competition, when
an official goes round to weigh each catch and award prizes
to the winners. The fish are put back in the water when
they have been weighed, to provide sport for other anglers
on another day. It may sound a delightful sport for the
anglers which entails no great harm to their quarry, but it
is not as simple as that. The prize goes to the chap who
has the heaviest total weight of fish, whether it is one
monster or a netful of midgets. Time, therefore, is the
essence of success and, once hooked, the object is to yank
the fish out as quickly as possible, disgorge it, pop it in the
keep net and get fishing again. Having a barbed hook
impaling its mouth can be no fun for the fish. When ripped
out with a disgorger by brute force and ignorance, insult

141

is added to the injury. Clumsy handling may damage the scales, which may be attacked by fungus disease, and the prospect of going through the same ordeal at the next competition can't add much to the joys of fishy life. The good bishop might have had second thoughts about the relative hazards of being fox or fish if he hadn't been obsessed with the Top People's sport of salmon fishing!

So far as the effects of the sport on the quality of the countryside go, they are neither black nor white. On the credit side, there is no doubt that the sheer weight of numbers – and votes – of men who are determined to preserve the quality of the water where they fish is of equal benefit to those who do and do not fish but love unspoilt countryside. Without opposition from fishermen there is no doubt that Water Boards and the Ministry of Agriculture would vandalise even more river banks to improve drainage and sluice the water ever more quickly to the sea.

From the point of view of wildlife conservation, fishermen probably do more harm than good. Their very presence in such numbers on the banks undoubtedly disturbs such shy creatures as otters, who are already under more pressure than they can stand. By the same token, streams of people along the banks, often trampling long grass and rushes, must disturb birds which nest close to water, such as warblers and reed buntings, kingfishers and dippers, dabchicks and coot, wild duck and moorhens. The more fanatical fishermen are not above getting the responsible majority a bad name by shooting herons and kingfishers, whose only crime is to be better at the job than their human competitors.

Coarse fishermen, in particular, do considerable damage accidentally. They use split lead shot to clip onto their line with pliers to ensure that the bait sinks as far as the float will allow it. These little shot are fiddling things to fit and, since time is all important in a competition, a great many of them are spilt. Some waterfowl, swans in particular, feed by 'dibbling' with their broad bills in the mud, feeling for solid food like snail shells or spilled grain, which they sieve and swallow, while the fine particles of mud are swilled away. Most birds digest their food by passing it through their gizzard, which is an extremely powerful muscly organ, in which particles of lead are lodged. Food that is swallowed is ground into small, digestible particles when it passes through the gizzard. Lead shot that is ingested from the mud is ground to tiny particles when it reaches the gizzard, and the swans subsequently die of lead poisoning. There has been a catastrophic slump in swan populations in the last twenty years and the Government has promised to ban the use of lead-shot fishing if voluntary action is not effective. It is understood that fishing authorities have now asked members to use alternative material for weights but the snag, in such wet government requests, is that a maverick minority defy the ban, to their advantage, while the responsible majority put themselves at a disadvantage by complying. When such steps are necessary, it would be far fairer for the government to get tough and apply and enforce legal sanctions to the small minority who get reasonable people a bad name in the country. Finally, hundreds of yards of broken fishing line, left in or near the water, ensnares and kills birds of many species.

Hand-reared fish are as tame and as reliant on food from their keepers as tame pheasants. When trout are suddenly set free in a fishery they surge up for the bait cast by unskilled fishermen, so they should be left for several months to acclimatise to their free state before being tempted by bait. Some clubs, however, put reared fish in one day and invite members to be unsporting enough to take them out the next. The Put-and-Take brigade are ranked right at the bottom of the piscine peck order. The snag is that, with such a status-symbol sport, many of the members do not have much skill and, even when there are fish for the catching so tame that they will swirl to the surface at any disturbance resembling a handful of feed pellets hitting the water, there are usually a few duffers who can't catch a fish. It is a wonderful opportunity for the water bailiff to have a little profitable fun. He listens patiently to their tale of woe, asks precisely where they tried and, wherever it was, he tells them he will show them a better (secret!) place next day. He chucks in a few even tamer fish before he takes them and, Eureka, the fish would leap at their fly if it was attached to a flat iron. The fishermen are overjoyed at their prowess – and the bailiff collects a very easy tip! But it does nothing to enrich the quality of the countryside!

The fact that leisure will certainly be a major growth industry in the next few decades makes it inevitable that the three and a half million fishermen will also increase in numbers over the same period. There should be no great difficulty in persuading the minority to abandon their lead shot, to pick up their discarded or broken fishing lines, which so frequently entangle birds, and to realise that leaving litter around and gates open are anti-social practices. The majority respect the countryside already. Fishermen are potentially such a powerful pressure group to resist bureaucratic damage to river banks, pollution of the water and unnecessary drainage of wetlands and marshes that conserve water, that they are invaluable allies for all who try to keep the countryside unspoiled. It may also be possible to persuade them to leave stretches of river – and more important, banks – entirely secluded. With such sanctuaries at intervals along the river banks, it might be possible for otters and shy birds to find enough safe habitat to survive without disturbance, for wildlife and Man to live in peaceful coexistence.

16

The Gentle Art

Having learned most of my fieldcraft the wrong side of the rural blanket, from patients of my father who regarded poaching as the perfect hobby for coal miners, I have the highest respect for such exponents of the art. They didn't poach for money and rarely sold what they caught. Times were very hard so that a rabbit in the oven was welcome at home and any surplus was usually given to their mates. The chief attraction was in pitting their skills against the skills of the gamekeepers, who were paid to defend the estate against such human predators.

Setting a long-net for rabbits across rough, tussocky parkland on a moonless night demanded the woodcraft of a Red Indian, and the collection of captured poachers' nets in the Squire's gun room bore eloquent testimony that trophies did not all end on the illegal side of the fence. To have any chance of success, they had to be competent naturalists, who knew the habits of their quarry so intimately that they could almost creep inside a rabbit's skull to read his intentions. They had to know every yard of the estate like the back of their hands, to evade capture if the keepers almost caught them in the act. Huge, muscly men slipped about with the silent stealth of tigers and, when the worst came to the worst, they could give a good account of themselves in a punch-up. But it *was* a punch-up with none of the armoury of knives and knuckledusters considered as standard equipment by modern tearaways.

I grew up with whippet men and chaps who had long dogs that were almost wise enough to talk. Like the legitimate shooting men of their time, they took more pride in their skills than the number of quarry in the bag and they poached not for profit but for pride and for the pot. Although I've never been a fishing man, I once made a tape recording which is amongst the pride of my possessions. The star was a Welsh miner who was ace of trumps amongst the salmon poachers and, since I was a dunce, in his eyes, who didn't know the ropes, he gave me a round by round description of his skills.

He began by asking if I could tickle trout, which is among the more basic of the piscine arts. You wade down shallow streams, muddying the water as you go, and the trout dive for safety under the overhang in the bank, cut by the flowing water. By sliding a hand gently along this shelf, it is possible to be feel the fish, which presumably imagines the contact to be nothing more than debris, floating downstream in the murky water. By gently tickling fingers along the fish's flanks, the gills can be felt and the sudden contraction of fingers into gills, pins the fish which can be lifted onto the bank.

When the subject of my interview had satisfied himself that he was not talking to a complete mug, he continued his account of his method with salmon. 'It's like tickling trout,' he said, 'only the shelf under the bank of swift rivers is far deeper than the lairs of trout in streams. You couldn't reach the back of it by probing at arm's length, so you jump into the water and take off your boots.' 'Don't you take your boots off first?' I asked. He looked at me with pity. 'No,' he said, and paused. 'You get in with your boots on and then, if the bailiff comes, you can tell him you've fallen in.' Such specialists know the places where salmon have been found lying before, so they don't waste time trying fruitlessly. When they come to a likely spot, the two of them jump in – and take their boots off. One grabs hold of a firm piece of bank and sticks his feet out horizontally, as if he was on the parallel bars in a gymnasium. He probes with his feet deep into the salmon's lie until he can locate a fish precisely. Then he takes a brass rabbit snare from his pocket and dives under, headfirst, while his partner holds him firmly by the ankles. It is vital to have his feet held because, he said, 'there is a jungle of "trash" (branches and other debris) under the bank where it is easy to get jammed by the current. Many a good man has been drowned like that,' he added, 'so you need a knowledgeable mate who will drag you out by the heels before it is too late.' If all goes well, the brass rabbit snare is slipped over the salmon and tightened where the body narrows to the tail. The poacher comes out, holding the rope on the snare in his hand, 'because,' he said, 'all hell is let loose for a bit.' It is vital to *hold* the rope, not loop it round your wrist, 'because, if the fish is a twenty-pounder and the water is deep, you may have to let go to avoid being dragged under.' Many a good man, I gathered, had also met his end like that!

Modern poaching, whether of fish or flesh, is no longer the romantic pastime it used to be. Salmon poachers are no longer loners or pairs, they go out in gangs with nets and sometimes with hand grenades or other explosives. They are not skilled artists either, interested in pitting their wits against the water bailiffs, but disreputable thieves as much bound up in profit from the numbers game as the worst type of syndicate shooters. If they get caught, they gang up and fight it out with clubs and knives and boots as ruthlessly as mobs of urban hooligans. They are as undesirable for their effect on the quality of the countryside as they are from the viewpoint of the sporting rights' owners, who have paid for the privilege, because they destroy salmon wholesale and, when they use poison, electric shock or explosives, they damage any other wildlife in the river.

Rivers are sometimes poisoned by tipping cyanide into the water, which 'knocks out' the fish downstream. They are scooped out quickly and thoroughly washed, which presumably renders the flesh safe to eat, since there are no prosecutions or reports of illness due to this cause.

In really remote rivers 'electric' fishing is carried out illegally, though it is common – and legal – in legitimate fisheries for coping with 'vermin', such as eels or predatory pike. The method is to pass high-voltage electricity between two spaced electrodes which stuns any fish passing between them, and these can be scooped out before they recover. Since a boat and electric gear are required, the poachers are obviously vulnerable to capture except in very remote stretches of river.

Such practices could best be thwarted if the market was controlled, because most of the outlets are hotels, who are prepared to pay a reduced price in cash for the fish and ask no questions about where they came from. It would certainly help if it was illegal to buy salmon except from a registered game dealer, who would lose his licence and his livelihood if he did not keep adequate records to prove that all the fish he sold had come from legitimate sources which owned or controlled the fishing rights. This is not as easy as it sounds because so many fishing rights' owners prefer unrecorded cash deals which do not come to the notice of the taxman.

The poachers of feathered game fall as far from the standards of old poachers 'for the pot' as gang poachers of fish for profit are from the boys who just nicked odd ones for sport. It is now common for downtown yobbos from big cities to sally forth on Sundays with spades and terriers to dig out badgers illegally and take them to sleazy pubs for the 'sport' (!!) of badger digging, which has revived since Ministry of Agriculture pest officers drew attention to the potential by gassing more than fifteen thousand. Badger digging is strictly illegal and conservationists are as keen as the police to have it stopped, but the Men from the Ministry have given it spurious respectability. Scores of prosecutions a year are taken out against badger diggers but, despite fines of several hundred pounds, the practice is still spreading.

Pheasant poachers often get on site just as dawn is breaking. They usually take a .22 rifle in the van with them, with which they have no difficulty knocking off a few brace of tame, hand-reared pheasants, which frequent the early-morning roadside verges in search of grit. These hit-and-run roadside raiders are extremely difficult to catch because they stop, do their execution and go so quickly that it is only luck if someone sees them. They will not be locals but complete strangers and, since they may well have fitted false number plates, they may be impossible to trace.

Rabbits are scarce in many parts, because of myxomatosis, and at best their value is only about a pound or little more, so that poaching them for profit is now a non-starter. In any case, farmers who have enough to do any damage are probably only too pleased to let any honest neighbour catch them for nothing.

The real Big Time poaching is for pheasants which before Christmas will probably sell for three or four pounds a brace. Shooting men have now discovered that it costs up to ten pounds apiece to rear them, when food and keepers' wages and capital equipment are counted, so that the loss of pheasants is far more than the profits which accrue from poaching them. Intensive husbandry means that plenty of medium-sized shoots release between 4,000 and 5,000 into the woods annually and, being very tame, they present no problems, even to urban poachers. Most frustrating of all is the modern practice of taking pheasant poults from the release pens as soon as they have been reared and put out in the wood, long before the season starts. They cannot fly properly, because their wings are trimmed before they are put out to prevent them straying before they settle down, and they are imprisoned in a wire-netting enclosure. They are simpler to catch than domestic hens and, instead of fetching a couple of pounds from the game dealer, unscrupulous shoots will buy them alive for twice the price, to stock their shoot with 'ready-made' birds to avoid the trouble of hatching and rearing them, asking no questions about their origin.

Temptation such as that, which can net more in a night than a major burglary, attracts criminals who bear no resemblance to conventional poachers. They turn up in gangs prepared to club or knife their way out of trouble and, while they are there, they will take anything else of value that is portable. They get out the relevant one-inch Ordnance maps from their local library and reconnoitre the patch they propose to poach by walking all the public footpaths, in broad daylight, as assiduously as the local Footpath Society, which they may have joined. One of the reasons that so many farmers and landowners detest public paths so much is that they are an open invitation to any crook who wishes to case their joint. And if shooting were to become either totally uneconomic or illegal, a great deal besides pheasants would disappear from the countryside because good game cover is equally attractive to a very wide selection of other birds and animals.

By far the most objectionable form of poaching, to my mind, is deer poaching. In the wild highlands of Scotland, a few red deer have always been poached by men with rifles, who are probably quite as competent as plenty of those who rent the stalking. If a deer is to be shot efficiently, and therefore humanely, I doubt if it would mind whether its execution is legal or illegal. A few are shot, at relatively close range, by men with shot-guns, designed for birds in flight. The shooter may be a farmer defending his crop or a poacher out for venison but, in either case, shot-guns are not acceptable because wounded animals are likely to die a prolonged and agonising death. Certain sizes of shot in shot-guns are legal for deer but I have seen frightful havoc wrought with them and would like to see them banned. The other reason for banning their use is that the police are *extremely* particular about granting permits to use a rifle and it has to be stated, among other things, where the rifle will be used with full permission. Since no one will have permission on the land he poaches (or it wouldn't be poaching) it would be possible to prosecute poachers for illegal use of firearms – which is a very serious offence – as well as conventional poaching, which is sometimes winked at by magistrates who are either left-wing or politically trendy.

Undesirable as poaching with guns or rifles may be, poaching with dogs is far worse. A great deal of afforested land is within walking distance of conurbations and an unacceptable aspect of the fashionable nostalgia for traditional rural values is the increased number of lurcher dogs. A lurcher, by definition, is a crossbred dog, deliberately bred so that three-quarters of its blood is of greyhound or whippet stock, while the remaining quarter was contributed by some other breed, specially selected to give definite qualities. Lurchers are usually produced by mating a half-bred greyhound or whippet back to pure-bred greyhound or whippet. The outcross can range from large deerhounds or collies, for the greyhound cross, to small terriers for whippets. Whippet lurchers are ideal for catching rabbits and greyhound lurchers will catch hares, or even deer, if crossed with other large breeds including pure-bred deerhounds.

When I was a lad, about the only people who kept lurchers were gypsies, or other travellers, and very occasionally poachers. They kept them for filling the pot with hares and rabbits, which they did extremely efficiently. After the last war, it became fashionable to run 'Lurcher and Hunt Terrier' shows, at which it was common to run lurcher races after artificial hares, usually a bundle of rags dragged up the track by a rope on a revolving drum, driven by an electric motor. Such races are now run at most Country Fairs. They are great fun to watch and very popular. Unfortunately, the type of urban hooligans who go badger digging became interested and discovered the pleasures of coursing hares with lurchers. They went out in gangs on Sunday mornings, often slipping several dogs at the hare, which was quickly mobbed, so they returned home with plenty of booty.

The gangs were large and rough enough to go where they liked without bothering about the formalities of getting permission. My neighbour, who farms a flat piece of land that was cleared for an aerodrome during the last war, saw a gang of yobbos in line abreast with lurchers, so he jumped on his tractor to investigate. When he got up to them they told him to clear off home or they'd pull him out of his seat and kick his head in! By the time he returned with the police, they had gone. This type of hooligan has graduated from hares to deer. They use big, strong deerhound-cross lurchers, often advertised as 'H, K & C. Good on deer'. For the uninitiated, H, K & C means that they will Hunt, Kill and Carry hares. Good on deer means exactly what it says.

A great deal of woodland, especially Forestry Commission woods, is varicose-veined with public footpaths. Poachers can prospect it to their hearts' content during the hours of daylight, with or without their dogs. They return after dark with powerful spotlights, with which they scan the glades and fields in and around the forests. Like most wildlife, deer do not associate artificial light with danger (or there would be less killed by motorcars). So the poachers can spot them and slip their dogs to course them. A good lurcher soon comes up to a deer and usually has no great difficulty in throwing and holding it. He cannot kill it though, so that the air is filled with the most pitiful screams until the poachers catch up on foot, and slit its throat. Of all the obscene forms of mangling death, catching deer with dogs, incapable of killing them quickly, is impossible to beat.

It would seem easy enough to catch a gang of men, wandering in woodland with conspicuous searchlights, accompanied by dogs whose quarry raise the echoes with their piteous cries. The fact is that they are very hard to catch. One forest, on the Welsh borders, is particularly badly troubled by such poachers. Forestry staff and the local police have worked out a scheme where one forestry warden sits on the highest hill with binoculars and walkie-talkie radio. He watches for the tell-tale lights, without which the lurchers cannot see their quarry, and he radios to the rest of his party where the gang is working. The police are past masters in finding their way round the winding lanes without lights and the forestry chaps know every track that will carry a Land Rover, so they try to surround their quarry. But it is easy to nip through thickets impassable to vehicles, leaving the trophy if really hard pressed. By the time police or staff do catch up, the gang is on one of the public paths, pointing out, in no uncertain terms that they are only taking the dog for a walk and why are they being persecuted? The fact that they may be a hundred miles from home 'taking the dog for a walk' or that it is three o'clock in the morning cuts no ice with them, since they are within their legal rights so long as they are on a public right of way and are carrying no incriminating evidence.

Such people do not eat the deer they catch themselves, they cash in by selling it for venison. As with salmon, the usual market is the hotel and catering trade, who ask no questions so long as the price is right. There has been strong pressure to issue numbered tags to people who can legally kill deer, so that the tag could be attached to the carcase and could be used to record the movement of the deer thereafter, to game dealer, hotel or whoever. It would then be an offence for any dealer or restaurateur to have a carcase that was not legally tagged on his premises, and the man to whom the tags were issued would be obliged to return them at the end of the season or name the people who had bought the venison. This would have the effect of limiting the killing of deer out of season and increase the risks of buying deer that had not been killed legally till it became uneconomic to take chances. Unfortunately, the same objections apply as with buying poached salmon. So many people are involved in the tax fiddles of cash deals that they would rather poaching continued than risk losing the tax-free profits of unrecorded deals.

17

A Cottage in the Country

The in-thing nowadays is to have a cottage in the country. Every chi-chi magazine is full of eye-catching ads extolling the virtues of any barn, without cavity walls or a damp course, which has been 'converted' to a 'desirable country cottage-style residence' that will fetch £40,000 north of Watford and double that in the 'civilised' south. Clapped-out rectories, with high rooms large enough to need a private power station to heat them, are described as desirable country residences. The rector lives in comfort in a villa with all mod. cons. Anything with more than an acre of land is a 'small estate', pools are 'lakes' and a couple of acres of scrub woodland is a 'sporting estate'. A derelict cottage, where the roof has caved in, is 'ripe for development'. In all too many cases, the estate agent's word mod. con. should not stand for convenience, but confidence trick.

The whole phenomenon has grown out of the instinct to escape from the shifty insecurity of the modern urban ratrace to what people believe are the more solid values of traditional rural life, where nothing ever changes. Lungfuls of clean air will be more healthy than the city smoke, home-grown spuds, dug fresh from the garden, will be so much more healthy – and appetising – than chips from the local take-away. Gossip in a village pub will be a refreshing change from endless efforts to convince acquaintances in city restaurants that your holiday abroad was far more glamorous, prestigious – and expensive – than theirs. Once the germ of an idea that a cottage in the country would restore a sense of security enters your mind, it will mushroom to an obsession.

The snag is that there are really two very different reasons for the overwhelming nostalgia for a solid rural life. Some people yearn to escape from the teeming, impersonal crowds in big cities to be integrated with a small community, where everyone knows his neighbour – though experience will soon show that by no means all neighbours like knowing each other! And, if they don't, it is far more difficult to ignore them or to escape from them than it is in a faceless city. The ambition to become involved with a village community and to make a meaningful contribution to activities there may well be frustrated because so many villagers have been bred and born in the same circle of friends and relatives that the community is self-sufficient – for every old inhabitant of a village seems to be related to all the others. Such close-knit communities do not always welcome strangers, especially if they throw their weight about. However well-intentioned advances are, it is wise to make haste slowly. Incomers are not accepted as locals for the first twenty years or so, however superficially courteous their reception is.

A quite different reason for wishing to 'escape' to the country is to enjoy the amenities of fresh air and wildlife, striding across green fields in sunny weather, revelling in peace and solitude, where it really is possible to get away from it all but at the same time retaining the comforts of civilised, sophisticated city life. Ideally, such people should settle for living in small country towns, which already have urban amenities, such as libraries and luncheon clubs and dramatic societies, squash courts and the rest. Those who enjoy such things will be welcomed, if only for the contributions fresh subscriptions bring, and all the attractions of living in the country will be on their doorstep. But when they settle in small village communities, the fresh blood and ideas they bring are often anything but welcome.

The usual pattern is that vote-carrying demand to settle in a favoured area spurs local authorities to concoct 'village plans' which involve very considerable expansion of the population. After a lifetime in a village of less than a thousand souls it is possible to know a very high proportion of other villagers, at least by sight, and strangers stick out like sore thumbs. However, the local authority authorises the building of additional houses to cope with, perhaps, an extra five hundred people, who will obviously make a serious contribution, via the rates, to running the village. The planners salve their consciences by declaring the original centre of the village a 'Conservation Area'. All this really means is that they will not give planning permission to alter the exterior shell of the original cottages so that strangers, passing casually through, are unaware of any alteration.

Planning permission for the new develoment is restricted to specified housing estates behind the main street, which brings an incurable itch to the palms of the local speculative builders. These people throw up a rash of new houses, using the latest cost-efficient techniques, such as putting a floor from wall, to wall of the external shell and putting the dividing walls, of hardboard and laths, directly off the central floor. By the time the main joists rot and cause internal room walls to crack, the mortgage will be paid and the owner can carry the can, while the builder retires to the Channel Isles. Many of these modern houses are mass-produced, as identical as the little boxes in the song, and they are so close together that there is no more privacy than there is in a suburb. But the inconvenience of commuting to work, especially in the winter, is far worse than jumping on a suburban train.

That is bad enough for those who genuinely want to be integrated into a small community, but the type who might have been happier in the more sophisticated atmosphere of a small country town are prone to take active steps to improve their lot. They feel sorry for people who have put up with being deprived of what they regard as the civilised necessities of life, so they join the local organisations to get matters improved. They are often young executives and their wives, who have got a cottage in the country as one of the outward and visible signs of an up-and-coming successful career. They are well used to the stresses of competitive life – and they have management skills, or they wouldn't be able to afford to buy a place in the country and commute to work. With the very best of intentions, they feel rather sorry for the swede-gnawing yokels who have put up with deprivation for all too long. Before anyone realises what is happening, they are on the Parish Council, the Parochial Church Council, and their women, who are equally accustomed to the executive rat-race, are queen bees of the local Women's Institute, Gardeners' Guild and Village Fête Committee. Indigenous villagers, courteous and shy as real countryfolk are, are outwardly polite but seethe inwardly. Street lamps and public lavatories which pop up as the result of the organising ability of the new-comers do not strike them as relief of age-old deprivation. Generations of locals have been begotten beneath discreet unlit front porches, as courting couples say good night, and they see no necessity for spotlighting their natural indiscretions. And nobody but strangers use a village public loo.

Some newcomers may go to extremes. They think that life is not complete without a Sports Complex for the community and, maybe, they would be right if they had settled in the local market town. But room for such amenities is limited in a village, where every space that owners were happy to do without has been snapped up to build houses or pub car parks or playgounds for the kids. The only available field in the centre of our village is owned by one of the oldest and most respected inhabitants who made it clear that she did not want to sell. At any price. The locals understood, but newcomers thought it was selfish and, before the old lady knew what had hit her, a compulsory purchase order landed through her letterbox. However quiet and inoffensive they may seem, countryfolk are not pushed about with impunity. She appealed against the order and won. In due course, the same tactics were tried again and, this time, the village was well and truly split, a village meeting was called and one of the District Council's Big Wheels arrived to settle the matter. The Anyone-for-Hockey? ladies were out in full cry and, since some of them were farmers' wives who had jumped on the newcomers' wagon, they were posed the pertinent question about why they didn't have a hockey field on the old man's farm? It was instructive to see how annoyed they got when the tables were turned. Right triumphed and the old lady still has her field intact, but the damage that was done to relationships between the old and new was incalculable.

It illustrated the fact that a sudden drastic infusion of new blood into old and traditional communities may alter the whole ethos of a village, though the arrival of quiet and sensitive strangers is welcomed with all the hospitality of real countryfolk. Some small villages where this has happened have been rejuvenated, so that old and new mix freely and happily to their mutual benefit. They have thriving societies and clubs, dances and functions which are as simple and unsophisticated as rural functions traditionally are. But the initiative has usually been left with the local inhabitants, who resent nothing more than being frog-marched into something that they are assured is 'progress' by strangers who regard themselves superior.

Like it or not, the countryside is changing due to the revolution in mechanised and intensive farming and there are fewer and fewer people working on the land as it becomes more and more mechanised. Although agricultural wages are still relatively low, they have improved enormously since the war so that fewer farm labourers live in tied cottages on the farm or even cottages in the village. Many have migrated to the nearest council housing estate so that they commute to work in cars. In common with other villagers who owned cottages, a great many have cashed-in on the Olde Worlde Cottage bonanza, sold theirs for more money than they ever dreamed of and prefer the modern houses built by the council, though it is doubtful if these will outlast the cottages they left.

Farm mechanisation has also resulted in many small farms being amalgamated because it is impossible to spread the cost of multi-thousand pound implements over the output of a small farm that does not really warrant such heavy capital expenditure. This has left many small farmhouses vacant and the most profitable course is usually to sell them, with a few acres to graze a pony, to rich commuters who can afford to modernise them. Planning authorities are naturally – and rightly – careful not to allow the old farmhouse appearance to be altered enough to spoil the rural character of the surroundings, though plenty of olde street lamps, carriage lamps each side of varnished oak front doors and similar unsuitable bric-a-brac are sprouting in most unlikely country places, and major difficulties occur when farmers, who do not want to dispose of homesteadings, find that traditional farm buildings are obsolete because they are useless for modern agricultural practice.

You can't get a combine harvester into an ancient implement shed, high enough only for wagons and hay wains; long, narrow cowsheds, built when cattle were handmilked, cannot be converted into modern, milking parlours; and graceful stable blocks, high enough for hunters and carthorses to be led through narrow doorways, are useless for modern tractors. Many farmers would like to demolish such out-of-date buildings and replace them with modern, concrete, purpose-built barns and grain stores and intensive cattle houses and, if they did, there would be a mighty outcry about despoiling the countryside. So councils may refuse permission to demolish them. Councils may also prevent major conversions of barns to houses, because they are anxious to avoid more housing – and therefore sewage and other services – except in villages.

It would surely be better to build the often obtrusive modern farm buildings either behind existing buildings or where they could be screened by existing high hedges or woods, when it is necessary to adapt to modern methods. Where farm buildings of pleasant character already exist, it seems unreasonable not to sanction their conversion to houses. Every barn converted would avoid a new building in the countryside and give the occupants the chance of privacy instead of being overlooked on a housing estate. If it is not practical to use existing farm buildings either for agricultural purposes or conversion to houses, why not sanction their conversion for light industry or labour intensive crafts? If the working rural population dwindles and is only replaced by commuters, many of whom do their shopping in towns where they work, the precarious fate of village shops will certainly be sealed.

Rural shops, bus services, public phone boxes and schools frequently go out of business because there are not enough people to support them. If planners will help to establish more rural light industry, there is still time to save them. The village where I live has two butchers, a post office, newsagent and village shop which sells anything from chocolate to cauliflowers, and wine to winegums. Although it is sponsored by an association of small shopkeepers, it is not in the same league as supermarkets when it comes to bulk buying. So it is not surprising that some of the prices are a copper or so above the nearest supermarket. They are also above whatever 'loss leaders' the supermarket is selling as a carrot for mokes who don't notice that the rest of their prices are often higher than the local shop. But our nearest town is six miles away and a return journey of twelve miles, even with a small car, will use a third of a gallon of petrol. You can buy quite a few items at a penny or so more, locally, before it pays to use sixty-pence worth of petrol!

The decline of village shops, because Big Business is putting them out of business, is very serious for village life because there are already so many commuters who own cars, and countryfolk who have had to find jobs in towns, that the demand for rural transport has slumped except at rush hour. It has reached the point, in many areas, where it is no longer economic for bus companies to maintain a service just for the few who wish to get to and from work on public transport and the dwindling number of elderly folk, who can't go shopping by car, would be bereft without the village shop. The commuters, who have migrated from towns to get out of the rat-race, can also find that they have landed up with the worst of both worlds. The only house available may well be on one of the modern estates the planners allow spec. builders to erect out of sight. The density of houses is likely to be about ten to the acre, so that neighbours can monitor each other's every move, from tumbling out of bed in the morning to creeping back at night. Privacy is as non-existent as it was in the suburbs and commuting to work may involve thirty or forty miles each way instead of less than half the distance. The fog and ice and snow of winter is incomparably worse than on well-lit roads, where the density of traffic prevents snow and ice clogging the highway. The disappearance of friendly, personal village shops can add the last straw by forcing housewives back to the big city for their weekly groceries. Here they will be lumbered with the very mass-produced, packaged mush that they had daydreamed would be a thing of the past when they got a place in the country. Village shops and some local transport are so essential to civilised rural life that a little extra outlay to

Lamping, or dazzling the deer with powerful lights and coursing them with lurchers, is one of the cruellest forms of poaching.

ABOVE Fishin': coarse fishermen contemplate their lines at Wollaton
Park, Nottingham.
TOP Shooting': driven pheasant shooting has become a status
symbol sport.
OPPOSITE Hunting': hounds are not the only enemies of foxes
which poach the keeper's pheasants.

TOP In order to encourage otters to repopulate waterways an otter holt is being constructed on the Somerset Levels.

ABOVE LEFT Village pubs seldom rely only on local trade any more.

RIGHT A disused railway line: symptom of the decline in public transport in the countryside, but which has been turned to advantage here as a nature reserve, Brotheridge Green Reserve in Worcestershire.

keep them viable is surely part of the price it is worth paying to keep the spirit of the countryside as attractive as its cosmetic appearance.

Our village, like many of similar size, has five pubs. That comes out as about a pub for three hundred souls. When we came, less than a quarter of a century ago, each had its regular clientele of local customers who not only knew each other well, but were so regular in habit that it was possible to predict who would be propping up which corner of each bar at which precise time. The pub that I used had three – and only three – types of solid refreshment. You could have ham, cheese or tinned salmon sandwiches – and you had to wait till nobody wanted a drink before the landlady would cut them. Now almost every village pub has pub food and most have Musak blaring from a slot machine which dispenses throbbing heads more surely and quickly than the most unwise surfeit of ale, and the bars are crammed with strangers, for village pubs now rely on casual trade from motorists passing through. If you want the local gossip you have to call at the paper shop or stop the postman. Saddest of all, most pubs have been tarted-up with chrome and leather bar stools and the landlords seem to change as often as football managers, each new incumbent passing on unmourned because he was not there long enough to give any continuity.

It is a far cry from the old days, when village publicans often owned the pub which they inherited from their parents, instead of being directed there as managers by breweries. In those days, every local had a character or two, skilled at conning free drinks from unsuspecting strangers. It was one of the chief amusements of the regulars to see demonstrated just how gullible the smart guys from far-off cities could be.

In quiet times, plots were hatched over foaming pints that controlled the running of the village far more efficiently than strangers elected to the council can do today. Now

163

there are few quiet times and too many interruptions! Pubs, I fear, will never be the same again.

Many people migrating to the country will surely be disappointed by the village church. Superficially it is probably as dignified and venerable and beautiful as ever it was. But in such times of insecurity, when no man's job is safe and the distant rumbling of the gods of war is never silent, it is not unnatural to expect the church to be the sheet anchor of solid, reliable, constant comfort that it has been down the centuries. It is not unnatural to expect the traditions that kept our sires and grandsires sane in troubled times to offer the same comfort to us. There is great solace in singing familiar hymns and hearing the familiar words that have never been excelled for the beauty of their prose. All too often it is not like that. In the old days, the parson was a man whom all countrymen respected. He was often the younger son of the family who owned the local great estate and quite likely most of the village too.

He will probably have been replaced by some trendy whipper-snapper, who calls parishioners by their Christian names on first acquaintance – and expects them to do the same with him. Being utterly unable to be better than – or even as good as – his predecessors, he will take every opportunity to be different. He will 'modernise' the service, translating the peerless prose to with-it jargon that owes nothing to tradition and gives no hint of continuity. This will probably get even worse when the incumbent departs and his successor is determined to be different too. Venerable bells, that have bathed the quiet countryside in heavenly music down the ages, will be melted for scrap and replaced by a carillon or electronic abomination, that all but belches jazz. Gravestones, which spelt out the history of the village for all to read, will be ripped up so that the churchyard can be shaved like a city park and the altar will be wheeled down into the body of the church. Instead of regarding it as the symbol of unattainable mystery at the east end of the church, the congregation will posture round it, shaking each other's hands because they are so equal. If the parson can appear even more equal than his flock, he may get an Oscar for the best performance. Small wonder that rural congregations are dwindling as fast or faster than in the deprived centres of big cities.

There it matters less because, in towns, there are usually several churches so that parishioners who do not like the vicar or service in one, can find an alternative not too far away. In a village there is only one church, take it or leave it. Worse still, so many people have grown disenchanted that it is no longer financially possible for one congregation to fund a parson so that one man is frequently in charge of several parishes. If you don't like the chap in your own village church, it is heavy odds that the same face will peer over the pulpit in neighbouring parishes so that you will be as badly burned there as scalded at home.

Luckily, village doctors have not followed their clerical brethren into such decrepitude. I gather that most urban doctors have clubbed together in group practices, operating from Medical Centres which are efficiently impersonal. Urban friends tell me that their doctors rarely visit them and that, if they are too unwell to crawl round to the centre, they are ill enough for hospital. When they do get to the centre, the receptionist extracts their card from her card index and, after an interminable wait, they are directed to whichever doctor in the group happens to be on duty. If they see the same doctor twice, they are very lucky. They are also very lucky if they are taken ill during the week, for modern urban doctors are on duty only during normal working hours. Patients who fall ill at weekends are likely to get a locum from the local relief service, and if he speaks the same language, it is more than they should expect.

My father was a doctor with a mining practice in the

limbo between town and countryside. He was on duty seven days a week and did not retire until he was seventy-eight, annually going for a week's course at the post-graduate medical school of his old hospital, to keep abreast of developments, until the year he retired. He believed that the modern National Health Service would turn medicine from a valued profession to a vulgar trade. The prediction, I am glad to say, has not yet come true in the country. Or not in my part of the country, anyway. We still have a doctor who is in a partnership of only three and would be grossly offended by any suggestion that he didn't know his patients personally. Time does not matter to such men because their work is their vocation and they take visiting patients at home as part of their job.

The difference, I suppose, is that town and country practices attract very different types. In a town practice, where all patients are within spitting distance, it is easy enough to fill the list with people who really are near enough to come to a central group surgery, so why should Mahomet go to their mountains? Especially as Mahomet himself may have a place in the country, from where he commutes in office hours. Such men will justify their impersonal approach on the grounds of cost-effective efficiency, which would delight the hearts of the management consultants who workstudy industrial operations for ergonomic automation. Our doctor's practice spans a sparsely populated area eight or ten miles across. Many of his elderly patients don't own cars, buses to the local town can be counted on the fingers of one hand and, between villages, they are non-existent. So if he didn't visit, his patients would go unattended and if ever he and his two equally professional partners are replaced by urban tradesmen, health in our neck of the woods will take a nasty knock.

One of the saddest sights in the countryside is the impersonal shell of a deserted village school. Our village still has a thriving primary school and a headmaster who realises

166

that no sort of society, from mice to men, can exist in harmony without a code of discipline tailored to its needs. So we don't suffer much from the modern plague of vandalism and still have country kids around who don't regard good manners as élitest or effete.

Sadly, such institutions are closing down all over the country and kids are taken miles, by bus, to be indoctrinated by trendies, in jeans and sloppy sweaters, who have declined from respected professionals to tradesmen who prefer the history of trades unions to religious education.

The best village school I have been privileged to know was the Countess of Stamford and Warrington Primary School at Enville, on the Staffordshire-Worcestershire border. Mary Sheward, the headmistress, was a formidable giantess, in mental stature if not in physical size, and she had a passionate belief in bringing the countryside to what lessons she couldn't take to the country.

The Big House stood opposite the school and in the grounds was a heronry of ten or twenty nests, which were clearly visible from the class room. Mary would ask the young children how far away they thought it was. Some said five miles, some five yards and, when she had established their differences of opinion, she made them walk there, counting every step.

The tall lads did it in a bit over four hundred and the little girls took five or six. Having established the undeniable fact that when a countryman says somewhere is a thousand yards away, on the assumption that a pace is a yard, you have to take it with a grain of salt, she let them push a wheel there, which did precisely a yard a revolution. An accurate count established that the heronry was about four hundred and forty yards from the classroom. By the time this has been checked accurately with a surveyor's chain, those kids had a firm idea of what a quarter of a mile actually looked like!

The next step was to allow them to play with a stopwatch, which always fascinates young children. When a heron rose from her nest, the watch was started and when she flew over the school room, it was stopped. The elapsed time was how long it took a heron to fly a quarter of a mile. Multiply that by four, and that was how long it took to fly a mile and it wasn't a big step to learn how to calculate how many miles they flew in the hour.

They kept annual records of the number of nests and charted them on the wall, so that when graphs went up or down, they equated it mentally with good and bad years in the heronry. Arithmetic *meant* something to those kids because they could visualise what figures meant. I wish I could say the same!

The tiniest tots, knee-high to grasshoppers, were taken for country walks and encouraged to play imaginative games in a huge hollow oak up on the common. When they were thoroughly involved with it, Mary asked them what stories they thought the ancient tree could tell. Some thought it was where foxes hid to escape from hounds and others, more romantic perhaps, thought it was where Shakespeare wrote his sonnets. Geography meant little to them till she ran a bunfight to raise the money to take them to the docks at Liverpool to see the ships come in. The captain of a whaler was so impressed with their manners and intelligence that he persuaded each member of his tough crew to 'adopt' a group of kids and write to them about their voyages. The kids wrote back and such a mutual respect developed that, when the boat was eventually scrapped, the ship's bell was sent to Enville to be the school bell.

The reputation of that school was such that it was filled not only with children whose parents could not afford a private school's fees but also with those who could. Local doctors and lawyers and other professional men could not find a prep school to rival it and the pupils had a better chance of ending up at Oxbridge than if they had graced a ruinously expensive emporium of learning.

Plenty of other village schools were just as good but the modern craze for mass-produced education, which drags everyone down to the level of the lowest, is closing them down, as Enville was closed after Mary Sheward had left. Now modern village kids are brought up as commuters, travelling miles to school as, one day, they will commute miles to work. They are educated in such amorphous masses that individuality is squeezed out before it can blossom and they grow up with urban minds and values. I was a rebel at school and I didn't like my school mates because we had too little in common. Nor did I respect my school gaffers because they were never really in control and were easy butts of our coarse humour. I have no illusions about who would have been in the saddle if I had been at Mary Sheward's school but I reckon that, if there is ever any truth in the old saying that school days are the happiest of your life, I should now be looking back with great affection.

Transport, small shops and pubs, real family doctors and respected clergy are far more vital to the quality of village life than sports complexes and coffee mornings. The extra cost of keeping such basics viable is a price well worth paying by anyone who values village life and is prepared to put himself out to get it.

18

Beyond our Control?

Although my soul is in the countryside, which I love in all its moods and seasons, I am a naturalist at heart, and my feeling for wildlife is deeper and more intense than it is for the most beautiful scenery, picturesque village or fertile farm. When the future of wildlife is threatened, my impulse is to attack the culprits, with no holds barred.

I make no excuse, therefore, for returning to the subject of the Ministry of Agriculture, Fisheries and Food, despite the fact that I referred to them, at length, in *No Badgers in My Wood* which was published five years ago. I believe that no government department, pressure group or sectional interest has caused more havoc to wildlife, from badgers to beetles, wildflowers to waterways, butterflies and birds than the men from the Ministry of Agriculture. It is true that they have been egged on by the National Farmers' Union, which is mainly controlled by the barley barons and other immense financial interests, caring for nothing but profit, and for years they have dreamed-up subsidies and incentives which have encouraged large farmers to produce a vast surplus of beef and butter and corn and milk, which have then been ruinously expensive to dispose of. Small and medium-sized farmers who live on their land and love it as deeply as – and far more personally than – any casual visitor, have been forced to follow suit or go up the spout. Their image has been tarnished till reasonable members of the public have adopted farmer-bashing as an obsession, while the real culprits are not the farmers but the bureaucrats who tempt them with easy money, if they succumb to their wiles, and devise systems of payment which bankrupt them if they don't. A succession of Ministers of Agriculture, too wet to control their own destructive department, have connived at actions that have virtually made the Common Market bankrupt and ruined the quality of the countryside in the process. The politicians, not farmers, are the niggers in the rural woodpile. When ministers themselves farm on a large scale, there is obviously a great temptation to devise schemes to line their own pockets.

The example of MAFF vandalism that affects me most is their disgraceful campaign against badgers. It is a subject that is close to my heart because I have been continuously involved with them for more than thirty years and claim to know something about them from personal experience. I reared my first cub in 1953. He lived for ten years and I described him in *Badgers at My Window* (Pelham Books 1969). Since then I have had badgers in my wood at home under continuous observation, except for the short period when the colony was destroyed, I believe, by a neighbouring keeper. Working so closely, for so long, with wild badgers in my own wood, where the observations I could keep were almost unlimited, I am naturally critical of some of the superficial observations of boffins and 'experts' who pontificate on the basis of a few hundred visits they make, to distant setts, when they can spare time from classes at school or university. MAFF's record with badgers and bovine TB is a monumental catalogue of callous incompetence.

The danger of humans contracting bovine TB from cattle was virtually abolished by the pasteurisation or sterilisation of milk, and it was decided to tackle the problem in cattle themselves, by the compulsory testing and slaughter of cattle which reacted. The subsequent upsurge of TB among

171

our own population is almost exclusively of the human strain and coincided with the post-war rise in immigration from countries where the disease is endemic. Bovine TB is rare among humans in this country. The Ministry's attempt to control the disease in cattle was fairly successful, except in certain areas, notably in the West Country, where it still persisted, and the Ministry vets failed to pinpoint the cause. This produced great pressure, by farmers whose herds were slaughtered, because they objected to losing cattle from a disease which they were told had been eradicated elsewhere. Eventually, a road casualty badger was picked up showing symptoms of the disease and the post-mortem confirmed suspicions. Grasping at the straw, they jumped to the conclusion that badgers were the vectors and set up experiments, not to discover the source by unbiased observations, but to prove the theory they had concocted. They inoculated healthy badgers with a massive dose of bovine TB and confined them with healthy calves. Badgers are extremely territorial and the badgers fought bitterly among themselves and tried so desperately to escape that the yard had to be concreted and the doors steel-plated to contain them. Open wounds and stress are notorious for activating TB and the stockmanship was so lax that the calves ate off the floor where the badgers were excreting. When a badger died, he was not even missed and he lay rotting under the straw till his corpse was too decomposed to examine by post mortem. Even under such disgraceful conditions, it was six months before the calves contracted the disease and such close contact for so long would obviously be most unlikely in the wild. On this scientifically suspect evidence, the men from the Ministry decided to exterminate all badgers in their control areas where breakdowns had occurred in cattle herds and there were any signs of infection in badgers.

They have never been able to explain, if the disease really is as transmissable between badgers and cattle as they pretend, why it was not endemic in badgers all over the country, when thirty per cent of cattle were infected before the eradication scheme. Or, if badgers were generally afflicted – and the Ministry men hadn't noticed – why is it now only found in isolated pockets? The Ministry have no answer but they decided, on this suspect 'evidence' of their laboratory experiment, to slaughter all badgers where there was a breakdown of TB in cattle. They began with a public demonstration showing how to control badgers by snaring, using self-lock snares which have now been made illegal on grounds of cruelty. It does not seem unreasonable to expect the men from the Ministry to have had sufficient practical experience to know this, but they were not even skilful enough to snare their quarry round the neck so that it would die by strangulation within a reasonable time. Badgers' heads are small and their necks taper, so the Ministry rat catchers set their snares in large loops to catch the badgers round the body. The more they struggled, the deeper the wire bit into the flesh and the self-lock mechanism, which works like the toggle on the guy rope of a tent, allows nothing to slack off when once it tightens. One lactating sow, caught round the body, had her udder split right open as the wire bit in relentlessly and a witness, Mrs Murray, was so revolted that she took out a private prosecution against the then Minister, Fred Peart, for 'cruel ill-treatment of a badger'. He got off on a technicality but the court awarded costs against him to indicate how deeply respectable people despise such conduct – whoever is personally responsible.

The demonstration left such a bad taste that the Ministry of Agriculture, Fisheries and Food abandoned snaring (at least in public!) and embarked on a massive campaign of gassing. This continued for seven years, cost more than a million pounds and exterminated upwards of fifteen thousand badgers. The same setts were gassed many times (one sett was gassed *twenty-four* times!). The fact that this was necessary indicates that, when attractive habitat is denuded of a species, it is repopulated from the periphery by animals seeking a living, where there is no competition, as surely as water finds its own level. These uncontrollable population movements can come from infected areas to clean areas, so that local extermination can easily be counter-productive.

Public concern about the cruelty and scale of the campaign built up such pressure that the Minister was forced to halt the procedure and order a public enquiry by his friend Lord Solly Zuckerman, an ageing professor who had made a reputation for being an astute scientist and wily politician back in Winston Churchill's day. His report was widely regarded as a whitewash operation, but it did recommend that, 'experts from the government's Chemical Defence Establishment (the germ warfare scientists at Porton Down) be called in to devise improvements in gassing procedures'. As his lordship wryly remarked, 'they are as well informed on the subject as any people in the country'. It is also well known that Porton Down scientists are not unduly sentimental nor squeamish, but Zuckerman's advice backfired. The Porton Down men had been ordered to find 'improvements in gassing techniques', not to make judgements on the ethics of the action. It gives a clue about just how brutal MAFF are when the Chemical Defence Department reported back that the method being used was unacceptably inhumane because the badgers took up to twenty-five minutes to die an agonising death. It was also an indication of the Ministry's disgracefully callous approach when they spent seven years killing fifteen thousand badgers without even bothering to take the elementary schoolboy precaution of checking that the method they were using was acceptably humane.

Peter Walker was forced to make a humiliating U-turn and ban the method his men had used for the last seven years. In future, badgers were to be caught alive, in cage traps and shot humanely. Only 'exceptional', trap-shy badgers would be snared, shot or netted. They continued cage-trapping into the spring of 1983, until another public outcry highlighted the fact that they were catching and shooting lactating sow badgers and leaving their orphaned cubs to starve. It was typical of the pitiless approach of MAFF but, once more, the pressure of public opinion forced them to desist from their uncivilised action. They promised to release any lactating sows caught in their cage traps though, when I challenged them to explain how they diagnosed a muddy lactating sow in a cage trap, on a wet dark night, they admitted it was impractical. So far as their assurance that only 'exceptional' trap-shy badgers would be snared or shot goes, it was soon obvious that their word was not worth the paper it was written on. I wrote to enquire how many badgers had been caught in my own county of Staffordshire in 1983 and was told fifty-two. Further enquiry elicited the information that, of these, thirty-two had been cage trapped and no less than twenty snared. So much for their promise about 'exceptional'! Not only did it make a mockery of their integrity but it also made a fool of the minister, who had been rash enough to vouch for them.

The whole campaign, based on unreliable circumstantial evidence, has been packed with cynical duplicity and arrogance. It has never been proved that badgers transmit TB in the field but, even if MAFF are right in their diagnosis, the harsh fact is that their cure is not working. The Ministry's annual report, *Badgers and Bovine Tuberculosis 1983* confessed that there had been more herd breakdowns in Control Areas, where MAFF had been exterminating badgers, in the last three years than in the previous four. In the rest of the country, the reverse is true, possibly because MAFF actions are counter-productive, not only by causing uncontrollable population movements, but also because some badgers always escape when a colony is harried. Stress caused could tip the scale, where the disease was dormant, and already infected animals could pass the disease on to fresh colonies where they settle.

MAFF have been proved ineffective by an increase of breakdowns where they interfere, so that the cure they prescribe may be positively dangerous. Their case is further weakened by a truly astonishing admission about the spread of brucellosis, which is supposed to have been eradicated in cattle. Hearing reports of an outbreak in the Leeds area, I wrote to MAFF and asked how many cases had been reported. I was told that there had been 'twenty-five *or so*' (my italics) herd breakdowns in the Leeds area in the past twelve *or* fifteen (my italics) months. The uncertainty about numbers and times proved to be due to the fact that the Ministry attributes the cause of the outbreaks to 'cull cow movements' which the Ministry were unable to trace, '*because farmers' and dealers' records were inadequate*' (my italics). It is impossible to over-emphasise the significance of this admission. Quite simply, it means that illegal movement of diseased cattle, which *are* discovered, may well be only the tip of an iceberg of illegal movements that never come to light. I therefore asked the Ministry why, if brucellosis could be transmitted by illegal movements they didn't know about and couldn't trace or control, the same does not apply to bovine TB? They were forced to admit that it could and that, 'TB caused by cattle movements amounts to ten per cent'!! (That they know about. The Leeds confession of their inability to quantify or control movement of diseased cows suggests what they don't know may be far higher and may infect both badgers *and* cattle in areas most heavily stricken.)

Following the clue myself, I discussed it with a friend who is a professional cattle dealer. He said that he was not surprised that there had been trouble in the Thornbury area of Gloucestershire (a main MAFF control area) because TB had always been rife there as long as he could remember. His theory is that dealers buy in Banbury, Gloucester, Hereford and South Wales markets and hold the cattle until the price is right for resale. He added that experienced dealers could spot 'a stinker' in a bunch as quickly as the boffins in MAFF and, when they did, they swapped it with one in another bunch and bundled it off! Perhaps the Ministry should hunt rogue cattle dealers instead of badgers?

It is widely known that TB is a disease which is commonly found under conditions of overpopulation or stress or bad husbandry, whether among cattle or badgers or men. A diligent search under such conditions is likely to uncover cases in any species so that, if MAFF like to squander even more taxpayers' money on their witch hunts, they are likely to unearth victims. It is, however, utterly impractical and politically unacceptable to exterminate badgers, even if MAFF pretend that it might be scientifically desirable. And to continue exterminating localised pockets, as at present, has been shown to be counterproductive.

The obvious constructive alternative is immunisation, which MAFF will not accept with cattle because, once immunised, the animal cannot be tested as a reactor. The ideal alternative would be to immunise badgers. Specialists in human TB tell me that it would be very difficult to devise an oral vaccine that would not be destroyed during digestion before it had taken effect. It seems, however, that there is no theoretical difficulty about producing an injectable vaccine similar to BCG, which is commonly used for humans. It would then be possible, with such a vaccine, to cage-trap badgers in the vicinity of herd breakdowns, immunise them and release them where they were caught. They would then be harmless and would prevent infiltration by badgers from possibly infected areas by their normal territorial aggression. Even MAFF cannot deny that badgers have a natural resistance to TB – or the species would have become extinct after the first outbreak spread it. Some form of inoculation, to enhance the natural immunity in areas of special risk (to badgers and cattle) seems a far more constructive solution than the Ministry's present bankrupt bungling. No such vaccine is as yet available, but a scientist at the School of Pathology at the Middlesex hospital is working on the project with a PhD assistant. No government funds are available - so his first assistant was subsidised by Iraq. The Ministry team, who have failed so abysmally should be included in government spending cuts and the savings made should be diverted into constructive research. Vets of such low calibre would do less damage if they stuck to castrating cats!

An example of particularly offensive bureaucracy occurred at Folkington, near Eastbourne, in the summer of 1984. The area was densely populated with badgers, which had not been disturbed, and MAFF were anxious to examine a range of other mammals for signs of bovine TB. They trapped for three years and found signs of TB 'in badger droppings', though it is unclear whether they took badgers as well, without admitting it. On this 'evidence', they came to the conclusion that badgers were the reservoir (though there had not been a breakdown in cattle for about three years!). They decided to mount a major extermination campaign and called on the landowner and *informed* her that they were about to enter her land to trap badgers, also *informing* her that she had no right to question their authority or withhold her consent. She was elderly and powerless against their dictatorial arrogance and they said no private citizen could thwart the wishes of a government department. They brought cage traps, baited them without setting them for a period and them commenced to trap in earnest. Meanwhile, local animal-rights activists heard about it, obtained the owner's consent to camp on her land, and tried to disrupt operations, going to the extent of cutting up the traps with bolt cutters. I believe that such extreme measures can alienate public opinion although the dictatorial attitude of the men from the Ministry was quite indefensible.

The landowner contacted me for advice and I told her that MAFF have no right to interfere with badgers on private property without first getting a Control Order or the landowner's or occupier's permission. So I suggested she instruct her agent to tell them to get lost until they obtained a statutory Control Order. The advantage of this is that before a Control Order is operative, it must lie 'on the table' of the House of Commons for twenty-one days and if, during that time, any MP raises a prayer against it, the government must give time for a debate. The matter is so contentious and Ministry methods are so obscenely cruel that the widest publicity could do nothing but good.

The trappers caught about fifty badgers in the first few days and took them away to the laboratory for experiments to be carried out. Some of the cubs were not weaned and there was no way of keeping captured animals in their own social groups because there was no way of knowing which group they came from. Lactating sows in the laboratory could have left unweaned cubs back home to starve and there was obvious likelihood of major aggression if strange badgers were caged together in the laboratory. Operations ceased as suddenly as they started, either while MAFF checked on the legality of continuing without a Control Order or because of the hassle caused by the demonstrators. A couple of weeks later another gang arrived from MAFF, and the *gauleiters informed* the owner that they were going to trap and snare – and there was nothing she could do about it because her tenant, the occupier, had agreed. So they rode arrogantly roughshod over her.

Meanwhile, I had written-up the episode in the *Field*, which was a pioneer of opposition to Aldrin, Dieldrin and the other chlorinated hydrocarbons which MAFF recommended as a pesticide, which had such a devastating effect on wildlife.

Feeble attempts to give the Ministry campaign some shell of respectability were made by the appointment of a 'Consultative Panel on Badgers and Tuberculosis'. This was set up in 1975 to keep under review, amongst other things, 'the operations undertaken by the Ministry in order to eradicate Bovine TB from badgers and to monitor its existence in the badger population'.

It consisted, apart from dominant officials of the Ministry, of a hotchpotch of members, ranging from those with animal welfare at heart, such as the RSPCA, Universities Federation for Animal Welfare and Nature Conservancy Council, to the Country Landowners' Association and Agricultural Trades Union. One ex-school teacher member starred in a Sunday paper as a ' "world expert" on badgers, who claimed to have put badgers on the map'. Successful action to keep them there would be more than welcome!

So far, I have seen little evidence that the Consultative Panel has done much to defend badgers against local extermination by MAFF and even Lord Zuckerman didn't seem to have much confidence in them because he recommended, in his report *Badgers, Cattle and Tuberculosis*, that at the end of three years from October 1980 an overall review should be conducted and its results published, with its focus on changes in the prevalence of tuberculosis in badgers, as well as on the numbers of herd breakdowns.

The review body was only appointed at the end, not of three but of four years, and consisted of a professor of zoology from Aberdeen University as Chairman, an agricultural economist from Exeter University, and a vet. The Ministry's 1984 report was just out, showing that results in control areas were as derisory as usual, and that bovine TB was *not* a significant cause of mortality in badgers. The whole tragedy has set the taxpayers back well over a million pounds, so the agricultural economist could not have had much difficulty assessing whether it was cost-effective; herd breakdowns in areas where the Ministry was interfering in the West Country are still deplorable, so that the vet would need to scratch his head to justify it. And research on badgers indicates that, contrary to MAFF assertions, badgers *do* have a degree of natural immunity so that indiscriminate slaughter may only have resulted in killing off the potentially immune specimens.

This time the Ministry will not be entirely judge, jury and prosecution. A body called Wildlife Link has been formed under the banner of the World Wildlife Fund – UK, and has submitted extremely detailed criticisms and suggestions to the review body so that, at least, they will hear both sides of the argument. Lord Melchett is closely involved with this Badger Group report so that if MAFF tries, once more, to sweep it under the carpet or to blind its opponents with science, they can be assured that the matter will be thoroughly aired in the House of Lords.

A most disturbing fringe danger of the Ministry's cage-trapping exercise emerged in the autumn of 1984. Reports began to circulate about unauthorised cage trapping, thought to be to obtain badgers for the illegal and obscenely cruel 'sport' of badger baiting, first in Staffordshire and then in the West Country, near Cirencester. Actual trappers were not identified but the traps being used appeared identical with those used by the Ministry.

Enquiries at the local MAFF office in Staffordshire brought the reply, 'badgers are not currently being trapped in Staffordshire . . . At least, not by us'! A letter from Tolworth Tower, at Surbiton, to whom awkward questions are normally passed, said that the only 'official' live trapping in Staffordshire was in the Ipstones area. They couldn't even agree between themselves whether they were or were not trapping 'officially' – but they did agree that some traps had gone missing so that whoever had taken them could be using them to catch badgers for baiting! To spend taxpayers' money on research into an efficient method and bait for catching badgers, simple enough for even MAFF ratcatchers to use, and then to be so lax that it falls into the hands of illegal badger baiters, is running true to form.

The spokesman from Surbiton also admitted that, 'if there was an urgent need, we would loan equipment when a licence was issued to take badgers for the purpose of preventing serious damage to land, crops, etc.' He did not specify what 'etc' might cover or what, if any, stipulations were applied to the type of person licensed, the subsequent fate of the badgers or what safeguards were to be applied.

When the Ministry of Agriculture shows interest in any species of wildlife, from ladybirds to moles or spiders to rabbits, it is literally the kiss of death, not only for the boffins' target species but for innocent species which are killed by accident as well. When rats became immune to the anti-coagulent poison Warfarin, Ministry ratcatchers, too unskilled to use the methods of their more professional predecessors, had to be issued with other poisons, since scattering satchels of loaded bait was about the limit of their ability, as it is difficult to recruit any but the lowest grade of labour for such work. When rats are poisoned by acute poisons, their fellows soon get the message and become bait-shy, so Ministry scientists select poisons that kill too slowly for the animal to be able to associate the cause of eating the bait with the effect of dying from it. The question of whether or not it kills humanely does not cross their minds, since compassion is an emotion they despise as sentimental. While rats or other rodents are dying a slow death, they are obviously easier to catch than their alert and active fellows – and Nature designed predators to catch the weakest of a species so that the fittest would survive. It follows therefore that hawks and owls and cats are likely to catch rats and mice, poisoned by

MAFF ratcatchers, before the healthy ones, with the result that many barn owls, in particular, died of secondary poisoning, carried by their dying prey. Old-fashioned farmers erected owl cotes in their barns to encourage barn owls, which are among the most successful controls of farmyard rats and mice, but a recent census shows that their numbers have halved in the last decade. The population of this most beautiful and useful bird is now at such a low ebb that there is serious concern for the survival of their species. This danger of killing innocent with guilty by secondary poisoning is so great that there are even a few members of Ministry staff who are responsible enough to worry about it, though there is little they can do directly.

When lactating sow badgers were being killed by poison gas and then cage traps, leaving their orphan cubs to starve, I was told unofficially by his colleague that one Ministry vet, who had to do a post mortem on a dead badger, reported on the official records words to the effect, 'One sow' badger examined. No sign of bovine TB. Placental scars indicate that she had a litter this season. She' was still lactating. Orphan cubs presumably left to starve.' I was so disgusted by this rumour that I challenged the Minister to confirm or deny it officially. There was no comment, so I draw my own conclusions.

Another strong rumour was that, since myxomatosis now only kills about forty per cent of rabbits affected, the Ministry is working on methods of 'controlling' (!) rabbits with poison bait. Any bait that rabbits are likely to take will probably be as attractive to deer and hares, domestic stock and probably birds. So I arranged for the question to be raised in Parliament.

Peggy Fenner, the parliamentary spokesperson for the Ministry of Agriculture, is a past mistress at the evasive answer which can be taken as an accolade by those charged with answering parliamentary questions. She gave her assurance that no field trials were taking place on poisoning rabbits, nor had the Minister any intention of initiating any. She was, of course, completely truthful. What she did not say, because the MP did not probe, was that field trials *had* taken place on *baits* that rabbits would eat – which happened to be carrot! There would be no subsequent difficulty about lacing them with poison, such as a carbamate, when the time came to mount a campaign.

This was confirmed in August 1984, when a MAFF spokesman announced on radio that myxomatosis no longer controlled rabbits and, as a result of experiments, it would probably be necessary to mount a campaign of poisoning rabbits on a large scale.

MAFF tried poisoning wood pigeons with alpha chloralose with disastrous results, though in that case they claimed alpha chloralose was not a poison but a narcotic. They were unable to explain the difference between a fatal overdose of a narcotic and a lethal dose of a poison! So, if Ms Fenner's parliamentary reply about research on poison for rabbits was not deliberately misleading, perhaps she had misunderstood her staff or been misinformed. This possibility is all the more worrying because there are persistent rumours that MAFF is working on a poison for badgers to replace gassing as a mass exterminant.

Some techniques devised by Ministry vets and chemists are so abhorrent that many Ministry staff are so distressed that there is never much difficulty about getting inside 'leaks' that are worth following up. When I wrote *No Badgers in My Wood*, in 1979, the Ministry policy – 'if we get an outbreak of rabies in this country' is to take a square round it of twenty-four miles by twenty-four miles and lay poison bait to kill the foxes. About eighty per cent of cases of rabies on the Continent are spread by foxes, and some countries have been successful in laying bait containing vaccine instead of poison. Local foxes pick it up and are immunised, so as to be harmless, and the treatment does not cause the uncontrolled population movements that the vacuum, caused by poisoning, would do. Our boffins don't like the continental method because 'live' vaccine is used, which is thought unsafe here. As one would expect, MAFF prefers extermination, so, when I wrote the book, their proposal was to lay 8,000 poultry heads, laced with strychnine, in the control area. This was unacceptable to anyone in his right mind because strychnine is one of the most acute and persistent poisons known to man. It kills the victim and whatever eats the victim and whatever eats that up to five times, it is said.

So I made the plea that, while there was yet time, they should do research to find a) a less persistant and acute poison and b) a bait that foxes would eat but dogs and as many other species as possible would reject. Although nothing was said officially, it soon came down the grapevine that scientists at the Chemical Defence Department, at Porton Down, were working on the project in their spare time from germ welfare. Early in 1982, I heard that MAFF had conducted field trials on an undisclosed Army firing range, using poisoned poultry heads and that, out of fifty heads used, the deadly poison in a large number had disappeared and could not be accounted for by its presence in victims of the bait. The explanation was said to be that the bait was too acute to handle safely, so had been encapsulated into coated pills, which could be inserted into poultry heads, readily obtainable in quantity from the poultry trade.

The theory for the disappearance was fascinating. Early in the year frogs and toads were spawning and are often eaten by crows and herons. It is obvious that their tiny ovaries cannot hold the large basinful of spawn each animal lays, so the eggs do not swell into gelatinous spawn until they come in contact with water after they have been laid. To avoid severe indigestion, the crow never eats the ovary but dissects its prey and discards the small sac that contains the spawn without eating it. The reason for the disappearance of the bait in trials may be that crows, not foxes, had found some of the fowl heads, dissected them and discarded the encapsulated bait because it resembled a frog's ovary. I was naturally interested to verify this but could not ask too pointed questions until the vet who was my source had left the Ministry.

On 1 December 1982, Ivan Lawrence, my MP, asked the Minister, 'What poison he intends to use against foxes, in the event of an outbreak of rabies; what trials have been conducted; how many foxes were recovered; and what other species of animals or birds were poisoned?' Ms Fenner confirmed that the Chemical Defence Department had been commissioned to produce an alternative to strychnine based on carbamate, which acts on the brain and degrades quickly; the trial was held on Ministry of Defence land at Kirkudbright; nine foxes were found dead (I had heard seven) and there was some evidence of small rodents taking the bait; baits not used were removed from the trial area. The 'small rodents', not surprisingly, were not specified! It may be significant that Ms Fenner did *not* mention that no birds were found dead, for I had heard that two crows were. It may also be significant that, although she confirmed all *baits* were recovered, she omitted to say that they were all complete with their poison capsules. It makes me wonder.

In any case, the prospect of 8,000 such dangerous baits, in fowl heads which dogs and cats and many other species would relish or children might pick up, being strewn around the countryside, is horrifying, though typical of Ministry callousness. Whether they followed the plea in *No Badgers in My Wood*, or whether experiments with alternatives to strychnine were coincidental is of small importance. (The book is mentioned in the Ministry's Zuckerman Report on *Badgers and Bovine TB* so I know they read it!) Whatever the reason for their experiment, I beg of them to find a bait that is more specifically chosen by foxes than other animals before they strew the countryside with thousands of doses of their foul poison. If they are short of practical experience with foxes in the field, I suggest they try rats or grey squirrels to hold the bait. Dogs won't eat them, but will roll on them when 'high'. Crows and magpies will take them but few mammals will. At least it would be marginally less dangerous than their proposals so far.

MAFF's obsession with poisons really is one of the major causes of conflict in the countryside and an important reason for the current farmer-bashing campaign, for which the blame lies not with farmers but with the officials who advise them. Every year the Ministry publishes a book under the title of *Approved Guide for Farmers and Growers*. It is nothing short of a catalogue of death. The Ministry of Agriculture has for years lived in the cloud-cuckoo world where prices do not matter. Not for them the old-fashioned dogma that there is any relation between Supply and Demand, by which real farmers in the real world have been ruled for generations. They are into quotas, ruled by mandate, percentage allocations, intervention prices and when Jopling, the current Minister, was almost drowned in the milk lake, he defended cutting our quota more than the French, 'because their milk yields are much less than ours'. Having spent years encouraging our farmers to get maximum production, at any price, his department was bulldozed into accepting the worst condition in the Common Market for them.

One way MAFF has seduced them to produce food we can't sell, is by the use of poisonous herbicides, fungicides and pesticides, to make a monoculture of whatever crop was the target. This dreadful book lists precautions necessary for each goodie on their menu. Stock has to be kept away from sprayed areas for up to six weeks. How wildlife are persuaded to take safe sanctuary is a minor matter of no importance to the Ministry. Such callousness spreads, and figures of birds found poisoned, which are reported to the Royal Society for the Protection of Birds, indicate that thousands are poisoned annually by accidental contact with sprayed crops, but hundreds are poisoned deliberately by bait laced with pesticides. Eagles and buzzards and kites – all rare – die from eating poisoned carcases laid specifically (and illegally) for foxes. The catalogue is endless, and I dealt with it more fully in *No Badgers in My Wood*, but things have not improved in the last five years.

The point I now make is that it is quite useless to expect responsible action from the Ministry of Agriculture. They have subsidised the grubbing out of hedgerows more than is vital to accommodate modern farm machinery; they have subsidised the draining of wetlands, ploughing of moorland and removal of small woods. All this has not been done because we need the food but to pander to the false god of maximum production, *at any price*, which taxpayers have had to subsidise. Success in this over-production has resulted in lakes and mountains of food we can't sell, so that, too has been subsidised to get rid of it. Don't blame farmers for following MAFF's advice and cashing-in on their crazy bonanza, blame the Ministry. The time has surely come to cut their corns, and the first major change I would advocate is that they should no longer be allowed to give official approval to *any* poisonous product. By all means let them recommend it, but let the decision whether it shall be approved – or for how long – rest with the Nature Conservancy Council.

The NCC could then evaluate the toxicity and attraction of the poisons and, starting with the worst, inform the agro-chemical industry that approval for their use will be limited to, say, two or three years. After that their sale will be illegal. This will provide manufacturers with the incentive to concoct *safer* rather than more speedily acting agricultural chemicals. They will then compete with each other to supply the most selective and least persistent chemicals that will provide effective control. It is impossible to turn back the clock, and chemical pesticides are here to stay. Whether their side effects are more or less dangerous to wildlife and wild flowers depends on who controls their use and choice. Bitter experience has shown the Ministry of Agriculture to be criminally irresponsible, and there is no point in spending taxpayers' money on a Nature Conservancy Council that is not allowed to conserve nature. It would therefore be sensible to saddle them with the responsibility of clearing the mess that MAFF has made of the poisons farmers are encouraged to spew on the land. By the same token, the Nature Conservancy Council should control licences for clearing woodlands, draining wetlands and cutting and spraying wayside verges and 'controlling' protected birds. MAFF's responsibility should end with what it recommends. There is no point in protecting wild flowers, under the Wildlife and Countryside act, so that children can be prosecuted for picking a primrose, when the local authority sprays and mows verges in spring, obliterating wild flowers and ruining birds' nests. The Department of the Environment, a paticularly ineffective, toothless department, should be encouraged to earn its keep by giving it the responsibility for licensing – or refusing to license – schemes of drainage, ploughing of moorland and other projects recommended by MAFF which might affect the aesthetic aspect of the countryside.

19

My Patch

One day, with good luck and good management, the minority of maverick farmers will be cut down to size, the strident farmer-bashers will pipe down and the arrogant bureaucrats will be brought to book. Having said what I think about them, it is not unreasonable for me to put my money where my mouth is and cough up my share of the price of the countryside by doing something constructive as well as talking.

Just over twenty years ago, my wife and I settled in an isolated cottage, set on the edge of a ninety-acre wood. There is farmland on the northern, eastern and southern boundaries and a thousand-acre softwood to the west. Our cottage is at the south-west corner, looking out on to a paddock of an acre, behind which is an acre and a half pool at the edge of a clearing of about five acres. Near the house is a stand of almost forty acres of mixed hardwood, mostly oak and birch, but including quite a few alders, hornbeams and ash. Behind that is another forty-acre patch, which when we arrived had just been clear-felled and leased to the Forestry Commission for 999 years. The rent was half a crown an acre. At the north-east corner of the wood is an old, flooded marlpit, called Primrose Dell, and at the northern end of the hardwood by the house (Holly Covert) is a paddock of five acres called Daffodil Lawn.

It all sounds very idyllic – and well it may have been, in the prosperous days of the family who once owned the estate. But, when we arrived, it was overrun by trespass, and any primroses and daffodils rash enough to show their pretty heads were snatched out of the ground, to be carried off to wilt in jam jars. While we were negotiating to buy the place, we naturally came over quite often from our house ten miles away – and we rarely came without seeing some man or other, wandering round with a gun. There was a small heronry, of fourteen nests, in Holly Covert and it was not uncommon to disturb louts lobbing stones up into the nests, and I caught a local policeman's son damming the main ditch, so that he and his mates could bathe in the flood they created.

Holly Covert, the only standing woodland left by the tree fellers before our arrival, had no 'bottom'. The previous owner's herd of goats, said to originate from a gift from Richard II, had spent the intervening six centuries turning as much of the estate as possible into a desert, by ring-barking mature trees until they died. Seeds and suckers that tried to regenerate were grazed off to the ground. As a result, there were only two mature holly trees left in Holly Covert, though a contemporary account in the last century described a marvellous scene of forest-sized holly trees, encrusted in snow at Christmas-tide! All that was left, when we came, was a mat of feggy grass, which allowed falling seeds to sprout among its damp leaves – but let them shrivel because their roots could not reach down to the soil, though there were plenty of oak and birch.

The cottage was derelict, with no electricity and a single, stark cold water tap, that dripped into a sink that emptied water onto the garden because nobody had bothered to dig a drain. If you had the temerity to battle through the nettle patch, outside the back door, you eventually came to the cottage's only convenience. It was a 'two-seater' earth closet, which had to be emptied with a shovel. It was not as idyllic as cottages in the country are supposed to be and our friends thought we had gone out of our minds. But the

191

lack of aesthetic attractions, the low economic value of the wood, and the difficulty of commuting anywhere in foul weather ensured that it was cheap!

The great attraction for me was the potential there was for a wide variety of wildlife. I am no believer in the attractions of rarity for the sake of rarity because I believe that unavoidable changes in the countryside are making life difficult for a very wide variety of common species. It is more vital to provide sanctuary for them than to lay out fortunes for a few species simply because they are rare. I have never forgotten the excitement of waking to realise that we owned a bit of England, nor the challenge of trying to leave it richer in wildlife than it was when we found it. We wandered around at dawn and dusk, and the coarse belches and banshee wails of hungry heron chicks was music to our ears. Occasionally we came across a few terrified fallow deer, which had only survived so long by constantly outwitting the man we had so often seen with his gun. There were woodpeckers and nuthatches, redstarts and tree pipits, and nesting wood pigeons, endlessly crooning the laziest tune in Nature.

When we went into the wood at dusk, we often disturbed badgers foraging, but the most diligent search did not discover a sett where they could lie by day. The simple reason proved to be that our land is cold and clammy clay, inhospitable stuff to excavate. A field away, over my boundary, the soil changes to warm, well-drained sandy gravel, where generations of badgers had excavated a large sett. I had a tame, hand-reared badger for ten years before we moved here, so was overjoyed by the prospect of seeing far more of wild badgers than had ever been possible when I had had to journey from home to go 'badger watching'. Generations of sportsmen have learned far more about their quarry than academic boffins, who derive their knowledge from books and brief projects, designed to glean enough knowledge to write a superficial 'scientific' paper on some obscure facet of behaviour or physiology. So I followed the sportsmen's trick of building an artificial earth, or drain, which I hoped would be occupied at best by badgers, or by foxes for second choice.

Artificial setts, seem simplicity itself – simply a den, at the top of a slope, entered by two pipes of nine inches diameter, sloping gently down till they emerge through the side of the slope. The whole is covered with earth and turfed or seeded, partly as waterproofing and partly to blend into the surroundings. My first attempt was a flop. Thinking I knew better than people who had built fox drains for generations, I was determined to make a five-star model that would put conventional efforts in the shade. When I had laid the pipes and made the kennel, I covered them with a polythene sheet before adding the soil. Not a drop of water could seep through. I hadn't thought about ventilation and my sheet prevented the porous soil breathing. When I examined it to see why no tenant had come, the whole thing was fusty and musty with mould. It stank vilely and there was small wonder that it was spurned.

I have experimented with various types since and find that one of the best is a pile of large roots, bulldozed out, when timber is felled. An ordinary JCB digger is adequate for the job and when a pile of roots about as large in area as a hay rick, but not so high, has been pushed together, it is covered with soil about a foot deep and sown with grass seed. By laying a couple of tree trunks about a foot apart, each a foot in diameter, at the base before starting, a tunnel is made for tenants to creep to the centre of the pile and to explore and excavate further, wherever there is room. If the roots are not packed too densely, and an infilling of soil is added as the pile is built, animals have the opportunity to make a maze of tunnels among the roots, with dens in dry corners under large roots.

Rabbits usually occupy the pile first and, as they breed, the foxes enlarge the holes to their nests, discovering in the process the network of rabbit holes which are easy to enlarge. Badgers come last, when the whole thing has settled and dried out, but once they settle, they find the site ideal. From a practical viewpoint, such rootpiles are quite undiggable, so that raiding badger diggers will go away empty handed. They are also difficult to gas so that even when the Ministry of Agriculture ratcatchers were gassing badgers officially, they found root piles very difficult.

There is another method of deterring diggers, where badgers are in conventional natural setts, often started as rabbit warrens. When local farmers are replacing old wire-netting stock-proof fencing, the discarded netting is difficult to dispose of. If it is laid over a badger earth and covered with a few inches of soil, the grass grows through it quite quickly. Before this happens, holes should be nipped through the netting with wire cutters to coincide with the holes in the sett through which the badgers leave and enter. Ideally, the sett should be covered with several layers of old netting and, when the grass grows through it, the roots bind it inextricably together, it is as impossible to dig through with a spade as a spring mattress. Artificial drains can be made digger-proof by the same method.

So far as the rest of our wildlife reserve goes, I have used the same techniques as the gamekeepers of my youth, though adapted for different species. The first thing a keeper does is to provide his pheasants with privacy and seclusion. Like our own species, most wildlife thrives best on a sense of security; gamekeepers do not tolerate disturbance on their beats. So our first move was to remove the trespassers, which was a job my old Alsatian Tough enjoyed enormously. Her success was absolute. What is sauce for the goose is sauce for the gander.

Herons are among the shyest and most timid birds. They have to be because, although they are protected by law, there are too many trigger-happy louts who let fly at them

194

if the opportunity offers. So if anyone walks under a heronry when the birds are sitting on eggs, they leave their nests and settle at a distance they believe to be out of harm's way. When the danger recedes, the local carrion crows pluck up courage to return before the herons do – and swipe their eggs. Left undisturbed, the herons are perfectly capable of seeing raiding crows off. The proof of this pudding is that, as a direct result of neither allowing anyone else into the part of the wood where the herons are nesting, nor even going in ourselves while they have eggs or young chicks, the number of nests has increased steadily over the years. From the original fourteen, twenty years ago, it had increased to seventy-three in 1984.

I am not sure whether some of them have two broods or not. We count in the middle of May, before the leaves conceal the nests and, to make certain of our facts, we attach a garden label to each tree with a nest, showing the sequential number of the nest. Nests which have well-fledged young in May often have young in July. I believe these to be a second brood, but the bird books maintain that herons are single-brooded. High-power ornithologists, who go by the book, say that the first pair rear their brood and leave and that a second pair take the nest over and lay. They want me to allow them to ring the birds to decide the matter. The snag is that the early nesters – I believe old birds – nest about six weeks earlier than the young birds. So it is quite possible to have a nest containing herons ready to fly and a nest newly hatched in the same or adjacent trees. So, if a bird ringer climbed up to put rings on the young birds' legs, he might scare the old birds into jumping out of the nest and fluttering to the ground before they were strong enough to fly back. I do not think the knowledge gained important enough to risk damaging the birds – so I tell the ringers to get lost because the wood is a sanctuary for birds, not birdwatchers.

The second step a keeper takes is to control the predators,

which might harm his game. Fond as I am of pheasants fresh from the oven, I am fonder still of some predators which have enough difficulty in surviving without extra hazards posed by me. So I never touch hawks or owls – which are protected – or stoats or weasels, except in special circumstances. We live off the land as much as we can and have never bought an egg in forty-five years of married life. We also rear cockerels and ducks and geese for the table. Being a practical countryman, therefore, I have no scruples about trapping or shooting any stoats which threaten my poultry or pigeons. Weasels are rather different and have never done my stock any harm. A year or so ago, a weasel took to raiding the nest boxes I put up for birds. He got into a box right by the window, killed and ate the tit and young – and then discovered his belly was too large to get out of the entrance hole. I realised that, once he had discovered nest boxes provided such easy pickings, no box in the wood would be safe, so I took immediate steps to ensure he raided no more.

The wood is a major winter roost for carrion crows, jackdaws and rooks. The crows and magpies and jays, which also congregate here, are a menace to small birds, so I often leave a hunk of meat or other bait lying out in the paddock, while I am writing at my desk, and keep my rifle in reach. Wildlife populations are never balanced but fluctuate with weather and seasons and, above all, food supply. A loaded bird table attracts hordes of tits and other small birds, to feast on the surplus. A wood with an abnormal population of quarry species attracts abnormal numbers of predators.

The next step a keeper takes is to make sure there is plenty of food to keep his game at home. This may be grain that he throws onto straw-covered woodland rides, food he puts in hoppers or special crops grown to attract and hold his birds. So I do what I can to encourage as many berry-carrying shrubs and trees as possible to grow in the wood.

Hawthorn and mountain ash, blackberry and beech and oak are all encouraged. The forty acres that was rented by the Forestry Commission was planted by them with a monoculture of fast growing pines, which shaded out all undercover and provided no food. In view of the work I was doing on wildlife management, they relinquished the lease so that I could do what I liked with that part of the wood as well as hardwood Holly Covert. There had been a great deal of regeneration over the years, so I am gradually clearing most of the pines so that the wood can revert to natural hardwood. But I am not clearing large areas at a time. I look for good young oaks or beech or hornbeam or other useful – and attractive – hardwoods, and take out a circle of softwood around them to give them light and air and space to grow. Like keepers, I try to get the maximum 'holding cover' too. That is thickets of bramble or gorse or rushes, or undergrowth, where shy creatures can lie-up and hide in safety. The combination of peace and safety from predators, good food and good accommodation has done for my threatened species precisely what the same factors do for a keeper's game.

It is known that most woodland wildlife congregates near woodland edges and the greater the length of edge, the better the chances for wildlife. So I have cut a series of wide woodland rides, radial to my study window so that I can sit, like a spider in her web, monitoring the movement from one part of the wood to another. I can see, from my desk, what part of the wood a creature has gone to so that I know where to look, if I want to find it, or where not to go if I wish to avoid disturbance. Behind the pool, across the paddock by the house, I have cleared a glade of about four acres and sown it with grass and clover, to make an attractive deer lawn. We can sit in comfort and watch them through a pair of powerful glasses in the evening.

Some of the bare and open patches in the wood have been cleared by pigs. In olden times commoners, near forests, often had the right of pannage, which meant that they were entitled to turn pigs in at certain parts of the year. The pigs rooted and turned over the soil, as they searched for acorns and beechmast and other goodies. When they had done, they left a perfect tilth for seeds to grow so that regeneration was rich and varied. We have done this, penning the pigs in an acre or so of wood we need cleared and it works now as well as ever it did. The home-fed pork, from free-range pigs, is as far superior to supermarket mush as free-range eggs are to the pallid, rubbery rubbish from hens kept in a battery.

Wild flowers often do best on sward that is grazed and careful control of the population of the deer controls the density of grazing. Partly because of that – and partly because we never pick them – the density and variety of wild flowers has also improved.

But everything has not always been sweetness and light. For two or three years a syndicate took the neighbouring shoot who were copybook Flash Harrys. Their Bob-the-Killer keeper decimated my predators, including badgers, which did not understand the intricacies of maps and strayed over the boundary. Polite requests and objective argument were regarded as signs of weakness and got me nowhere. So I 'fed' Holly Covert hard and trapped, shot or drove in the wrong direction every pheasant I could, trying to make sure they had a one-way ticket. Unpleasant as it was, the stratagem worked and the shooters disappeared, to be replaced by a very pleasant lot with Andy, their most cooperative keeper. He and I get on fine. I let him come into my wood whenever he likes to drive the pheasants out. I never take a pheasant myself till the season is over and then I catch-up in 'live' traps. I put what cocks I want in deep freeze and give Andy the hens for his laying pens. For his part, he respects 'my' predators. It is a pleasant example of friendly give-and-take – but I have planted a lot of rhododendrons in Holly Covert, as well as encouraging

198

natural regeneration with pigs. As long as Andy stays it will provide marvellous cover for his birds – .which he can drive out as he likes. If he goes, I shall be arguing from a position of strength with his successor, because the wood could milk our end of their shoot of almost every pheasant. So I don't anticipate the same troubles in future.

We had a spot of bother with the hunt too, as I described in *No Badgers in My Wood*. Dermot Kelly, the first master, was a highly competent huntsman and, after I impounded three of his hounds which were chasing deer, we understood each other's strengths and never had another cross word. His successor was less competent and 'lost' his hounds in our wood one day, killing two deer and making the rest very nervous and spooky. Although he was very sorry, it was obvious he was not capable of avoiding a recurrence, so I put a deer- (and hound-) proof fence round the wood, which has cured such troubles for good.

My philosophy in managing my reserve is simple. Believing that the pressures of ordinary visitors to the countryside pose one of the greatest threats to wildlife, I have tried to demonstrate that it is as simple to encourage an unnaturally high population of wildlife species as gamekeepers find it to produce a high stock of pheasants. This population is likely to be more than the habitat can maintain. The surplus pheasants provide sport for sportsmen who enjoy shooting. My surplus wildlife spill over into the surrounding countryside, where visitors can enjoy seeing them without threatening their future.

So I feel that if planners could be persuaded to allocate small areas, of low economic value and low amenity value, as my wood is, they could form sanctuaries for wildlife, safe from the pressures of too many people. The specialised expertise that I use to encourage a high population of pressured species could be supplied by the members of any Naturalists' Trust or County Trust for Nature Conservation. All that they would need to supply a viable population of birds and animals and insects, for all to see, would be secluded areas near to National Parks or other places where so many people congregate that all shy creatures are put under pressure. The one snag is money? Not with me, it isn't. In order to prove my point – and because I could not afford it otherwise! – my reserve has to be self-supporting. My only capital equipment is an ancient 1957 Ferguson tractor, which cost £100 eighteen years ago, on which I have spent another £100 in the meantime. I have also got a chain-swipe (bought at a Government Surplus sale) to mow the rides and clearings, and a water roller, knocked up by our local blacksmith for a few pounds. If I want a new ride or clearing, I tot up the trees to come out, get quotations for what they will fetch for timber, and I only go ahead if the income will cover the contractor's cost of clearing the site and leaving it in a state for me to seed and roll.

Every year I sell a few hundred pine trees to contractors who cut them down, leaving the clearings I want round oaks and other hardwoods I wish to encourage in the wood. I sell the tops for Christmas trees and the poles for nutters to break their necks over when their horses jump over them – or don't! The cash I get for them is ploughed back to repair fences or clean ditches, or whatever I need doing.

To maintain a steady population of deer, it is necessary to cull about fifteen per cent per annum. Otherwise there would not be sufficient food to keep them and they would damage the wood. I like venison – and the surplus is sold to do other maintenance jobs. The reserve is self-supporting, so that any naturalists could do the same in other parts of the country, without any cost on the rates – if the planners would give them seclusion and freedom from disturbance.

When I tumble off my perch, I should be happy to be remembered as the chap who pioneered a chain of such reserves to contribute to the quality of the countryside – without adding to its price!

Index

Index